AN INTRODUCTION TO THE
STUDY OF ALGAE

AN INTRODUCTION TO
THE STUDY OF ALGAE

BY

V. J. CHAPMAN, M.A., Ph.D., F.L.S.

Fellow of Gonville and Caius College
University Demonstrator in
Botany at Cambridge

CAMBRIDGE
AT THE UNIVERSITY PRESS
1941

CAMBRIDGE
UNIVERSITY PRESS

University Printing House, Cambridge CB2 8BS, United Kingdom

Published in the United States of America by Cambridge University Press, New York

Cambridge University Press is part of the University of Cambridge.

It furthers the University's mission by disseminating knowledge in the pursuit of education, learning and research at the highest international levels of excellence.

www.cambridge.org
Information on this title: www.cambridge.org/9781107644014

© Cambridge University Press 1941

First published 1941
First paperback edition 2013

A catalogue record for this publication is available from the British Library

ISBN 978-1-107-64401-4 Paperback

CONTENTS

PREFACE

For a long time there has been a great need for a short and relatively elementary text-book on Phycology which would be suitable for University students, and also for those schools which include visits to marine biological stations as part of their curriculum. Such a text-book would not require to be too advanced and yet should survey the whole field of phycological knowledge, not only from the systematic but also from the physiological and ecological viewpoints. The two most recent works on Phycology do not entirely fulfil this function. Fritsch's *Structure and Reproduction of the Algae* must be regarded not only as a classic but also as a monumental piece of work, but it is somewhat unwieldy in size for the ordinary student and also it is a compendium of much that he does not require to know. At the same time it is a book that no University or research student in Phycology can afford to be without, whereas this present volume does not pretend to cater for the research student. The other work, Tilden's *Algae and their Life Relations*, is also somewhat bulky, and although it is perhaps more on the lines of the present volume, nevertheless it is primarily concerned with systematic phycology. It seems to the present author, therefore, that there is a place for a relatively short work on the outlines of Phycology containing the amount of information that could be conveyed in a course of lectures lasting over a period of 22–24 weeks at the rate of one lecture per week. No attempt has been made to produce any work more elaborate, primarily because Fritsch's volumes will fulfil that need. These, then, are the reasons for the appearance of this volume.

Relatively few types have been selected from out of each group; some of these have been described in considerable detail whilst others are mentioned merely to illustrate the course of development in either the vegetative or reproductive organs. Every type is fully illustrated because the present author firmly believes in this medium as the best means of teaching. Types that are regarded as essential for first and second year students are indicated by an asterisk, and even then it is not intended that they should necessarily absorb all the details about these species. It may come as a rude shock to some teachers to find that long-established friends, e.g. *Gonium, Vaucheria,* have not been asterisked. The present author believes that such types should have been omitted from curricula years ago either because they do not convey anything essentially new, or else because recent work has shown them to be wholly unsuitable types for elementary students. Up to the present, however, established tradition has kept them firmly ensconced in their position, but whether they will be able to retain it remains to be seen. It is suggested that third and fourth year students should study additional types selected from among the other species. Certain of the other chapters have also been marked as suitable for the first and second year students. Several chapters have been devoted to Ecology because the literature now available in this branch of the subject ought to be made accessible to the ordinary student. In these other chapters limits of space have rendered it necessary to select the material, and it may be felt by other teachers that some original work has been omitted that perhaps might have been inserted. In a book of this type such a feature is inevitable, and the author acknowledges that the choice of material has been a personal affair and that it is, as such, open to this criticism. There is a chapter on Physiology, Symbiosis and the Soil Algae, and also one that is devoted to a survey of reproduction and evolution. Part of one chapter is devoted to a brief account of the more important fossil types because it is essential that these should be studied and compared with their living successors, and also an acquaintance with these forms materially aids any discussion on evolution.

The algae are now divided into a number of groups, and whilst it is essential that the student should know that these groups exist, nevertheless, his attention should be concerned primarily with the major divisions. For this reason most attention has been given to the Chlorophyceae, Xanthophyceae, Cyanophyceae, Phaeophyceae, and Rhodophyceae. This is perhaps somewhat indefensible, but since the species which are normally encountered by the student belong principally to these groups, I believe the procedure is justified.

In order that the student should not be burdened unduly, only the more important papers have been provided in the references, but even these are appended only for those who are especially interested in the group. The bibliography therefore does not pretend to be complete, and the choice of what are regarded as important papers has lain with the author. It is sincerely hoped that the majority of University students will find all that they need to know for a degree course in the volume.

I should like to acknowledge the assistance that I have obtained from existing volumes and also my indebtedness to the following publishers for permission to reproduce figures from their works. Cambridge University Press: *The Structure and Reproduction of the Algae*, vol. I, by Fritsch; *Algae*, vol. I, by West; *A Treatise on the British Fresh-water Algae*, by West and Fritsch; Magraw Hill Book Co.: *Fresh Water Algae of the United States*, by G. M. Smith; University of Michigan Press: *The Marine Algae of the North Eastern Coast of the United States*, by W. R. Taylor; University of Minnesota Press: *The Algae and their Life Relations*, by Tilden; The British Museum: *Handbook of British Marine Seaweeds*, by L. Newton; University of California Press: *Chlorophyceae* and *Melanophyceae of the Pacific Coast of North America* (2 vols.), by Setchell and Gardiner; Dulau and Co.: *Phycological Memoirs*, by G. Murray.

In addition, reference was made to Oltmann's *Morphologie und Biologie der Algen* and to Kniep's *Die Sexualität der Niederen Pflanzen*, but owing to the present conditions it has not proved possible to get in touch with the publishers.

Acknowledgement for the use of figures is also made to the editors of the following journals, to whom my thanks are due: *Annals of Botany, Journal of Ecology, New Phytologist, Journal of Genetics, Botanical Gazette, American Journal of Botany, American Naturalist, Annals of the South African Museum, Transactions of the Royal Society of South Africa, Philosophical Transactions of the Royal Society, Transactions of the Royal Society of Edinburgh, Proceedings of the New Zealand Institute, Journal of the Linnean Society (Botany), Proceedings of the Linnean Society of New South Wales, Botanical Magazine (Tokyo), Bulletin of the Torrey Botanical Club, Publications of the Hartley Botanical Laboratory, Bulletin of the United States Department of Agriculture, Journal of the College of Agriculture, Tôhoku (Hokkaido) University, Reports of the Great Barrier Reef Expedition* and *Proceedings of the Cambridge Philosophical Society*.

x   PREFACE

Use has also been made of figures from the following periodicals
to the editors of which I tender thanks, although conditions have
made it impossible to get into touch with them: *Archiv für Protisten-
kunde, Zeitschrift für Botanik, Revue générale de botanique, Hedwigia,
Berichte der Deutschen Botanischen Gesellschaft, Jahrbücher für
Botanik, Planta, Flora, Beihefte zum Botanischen Zentralblatt, Le
Botaniste, Österreichische botanische Zeitschrift, Lunds Universitets
Årsskrift, Botaniska Notiser, Revue algologique, Protoplasma, Arkiv
för Botanik, Svensk botanisk Tidskrift, Nova Acta Regiae Societatis
scientiarum Uppsaliensis, Kongliga Svenska Vetenskapsakademiens
Handlingar, Kongliga Fysiografiska Sällskapets i Lund Förhandlingar.*
The authors of the various figures are acknowledged in the
legends, and references to the more important papers will be found
at the end of the appropriate section.

Much of this book has been inspired, and indeed used, during
class visits to the marine laboratories at Plymouth, Port Erin,
Lough Ine and Millport, and I can think of no better way of
becoming acquainted with the algae. These visits were initiated
under the tutelage of Mr A. G. Lowndes and to him must go much
of the credit for my interest in this branch of Botany.

I should also like to acknowledge gratefully the encouragement
and help given me by Professor F. T. Brooks, F.R.S., and Professor
F. E. Fritsch, F.R.S., whilst a special debt is due to Dr D. Catcheside
who read and criticized the whole manuscript. I am also indebted
to Dr H. Hamshaw Thomas, F.R.S., who read the section on Fossil
Algae, and to Dr G. C. Evans who read the chapter on Physiology
and the one on Ecological Factors. Dr Godwin also very kindly
read and criticized a portion of this volume whilst it was in
proof. Finally, there has been the help and encouragement given
me by my wife, and it is in no small measure due to her unsparing
help in the drawing of the figures and the preparing of the Index
that this book sees the light of day.

                                                        V. J. C.

Gonville and Caius College

May 1941

CHAPTER I

*CLASSIFICATION

In the older classifications the algae proper were simply divided into four principal groups, Chlorophyceae or green algae, Cyanophyceae or blue-green algae, Phaeophyceae or brown algae and Rhodophyceae or red algae. Now, however, that more is known about the simpler organisms which used not to be regarded as algae, it has been realized that there is no real justification for such a distinction, and so the number of algal groups has been increased. This is because it has become evident that the Flagellata and other simple unicellular organisms must properly be regarded as algae, even though of a very primitive kind. At present it is most convenient to divide the algae into ten classes, one of which, the Nematophyceae, is perhaps somewhat speculative. One of the principal bases of this classification is the difference in pigmentation, and a recent study of this problem shows that it is fully justified.

(1) CYANOPHYCEAE. The plants in this group show very little evidence of differentiation, containing only a very simple form of nuclear material, no proper chromatophore and no motile cells with cilia or flagellae. The products of photosynthesis are sugars and glycogen. The colour of the cells is commonly blue-green and hence their name, the colour being due to the varying proportions of the pigments phycocyanin and phycoerythrin. There is no known sexual reproduction, propagation taking place by simple division or else by vegetative means.

(2) CHLOROPHYCEAE. This group used to comprise four great subdivisions, the Isokontae (equal cilia), Stephanokontae (ringed cilia), Akontae (no cilia) and Heterokontae (unlike cilia). It is now more in keeping with our present knowledge to place the last section into a separate class, and this is the procedure adopted in most recent books. The plants of the Chlorophyceae exhibit a great range of structure from simple unicells to plants with a relatively complex organization, whilst the chromatophores also vary considerably in shape and size. The final product of photo-

synthesis is starch together with oil, and a starch sheath can often be demonstrated around the pyrenoids. In the bulk of the members of this class the motile cells are very similar and commonly possess either two or four flagellae, but in the Oedogoniales (Stephano-kontae) there is a ring of flagellae whilst in the Conjugales (Akontae) there are no organs of propulsion. Sexual reproduction is of common occurrence and ranges from isogamy to anisogamy and oogamy. The colour of the cells is usually a grass green because the pigments are the same as those present in the higher plants and, furthermore, they are present in much the same proportions.

(3) XANTHOPHYCEAE (Heterokontae). The plants in this group are usually of a simple nature, but their lines of development frequently show an interesting parallel or homoplasy with those observed in the preceding group (cf. p. 264). The chloroplast is yellow-green owing to an excess of xanthophyll, one of the four normal constituents of chlorophyll. Oil replaces starch as the normal storage material, the lack of starch being correlated with the absence or paucity of pyrenoids. The motile cells possess two unequal flagellae (occasionally only one) arising from the anterior end. Sexual reproduction is rare and when present is isogamous. The cell wall is frequently composed of two equal or unequal halves overlapping one another.

(4) CHRYSOPHYCEAE. These form another very primitive group in which the brown or orange colour of the chloroplasts is de-termined by the presence of accessory pigments such as phyco-chrysin. Most of the forms have no cell wall and hence are "flagellates" in the old sense of that term, although there are some members which do possess a cell wall and hence are "algal" in the old sense of the term. Fat and leucosin (a protein-like substance) are the usual forms of food storage, whilst another marked feature is the silicified cysts which generally have a small aperture that is closed by a special plug. The motile cells possess one, two or, more rarely, three equal flagellae attached at the front end, but in one subsection the paired flagellae are unequal in length. The most advanced habit known is that of a branched filament, e.g. *Phaeo-thamnion* (cf. p. 123), whilst the palmelloid types attain to a much higher state of differentiation, e.g. *Hydrurus* (cf. p. 123), than in either the Chlorophyceae or the Xanthophyceae. Sexual reproduction is not certain, and such records as there are point simply to isogamy.

(5) BACILLARIOPHYCEAE (Diatoms). One of the characteristics of these plants is their cell walls which are composed partly of silica and partly of pectic material. The wall is always in two halves and frequently ornamented with delicate markings, which are so fine that microscope manufacturers make use of them in order to determine the resolving power of their lenses. The chromatophores are yellow or golden brown containing, in addition to the usual pigments, accessory brown colouring materials whose nature is only just being established. One set of forms is radially symmetrical, the other bilaterally so. The presence of flagellate stages is highly probable in the former whilst there is a special type of sexual fusion in the latter group (cf. p. 122).

(6) CRYPTOPHYCEAE. There are usually two large parietal chloroplasts with diverse colours, though frequently of a brown shade, whilst the product of photosynthesis is starch or a closely related compound. The motile cells have two unequal flagellae and often possess a complex vacuolar system. Nearly all the members have a "flagellate" organization and there is no example of the filamentous habit. One type, however, has been described with a tendency towards the coccoid (non-motile unicell with a cell wall) habit, and so this must be regarded as the least "algal"-like class. Isogamy has been recorded for one species.

(7) DINOPHYCEAE. Most of the members of this class are motile unicells, but there has been an evolutionary tendency towards a sedentary existence and the development of short algal filaments, e.g. *Dinothrix* (cf. p. 126). Many are surrounded by an elaborate cellulose wall bearing sculptured plates and inside there are discoid chromatophores, dark yellow or brown in colour and containing a number of special pigments. The products of photosynthesis are starch and fat. The motile cells normally possess two furrows, one transverse and one longitudinal, although they may be absent in some of the more primitive members. The transverse flagellum lies in the former, and the latter is the starting point for the other flagellum which points backwards. Sexual reproduction, if it occurs, is isogamous, and it has not been clearly established in the few cases reported. Characteristic resting cysts are also produced by many of the forms.

(8) PHAEOPHYCEAE. This group comprises the common brown algae of the seashore and it is worth noting that the majority are

wholly confined to the sea. The brown colour is due to the presence of a pigment, fucoxanthin, which masks those other chlorophyll constituents which are present. The products of photosynthesis are alcohols, fats, polysaccharides and traces of simple sugars so that there is evidence of some diversity of metabolism. The simplest forms are filamentous, and there are all stages of development and increasing differentiation up to the large seaweeds of the Pacific and Arctic shores with their great size and complex internal and external differentiation. The motile reproductive cells, which possess two flagellae, one directed forwards and the other backwards, are commonly produced in special organs or sporangia that are either uni- or plurilocular. Sexual reproduction ranges from isogamy to oogamy, but in the latter case the ovum is normally liberated before fertilization. The life cycles may be extremely diverse and are perhaps better regarded as race cycles (cf. p. 246).

(9) RHODOPHYCEAE. The members of this class form the red seaweeds, and although most of them are marine nevertheless a few are fresh-water. Their colour, red or bluish, is caused by the presence of the pigments phycoerythrin and phycocyanin, whilst the product of photosynthesis is a material known as "floridean starch". Reproductive stages with locomotor appendages are not known, even the male reproductive body being without any organ of locomotion. The simplest members are filamentous, and again all stages of differentiation up to a complex body can be found, although they do not develop to quite the same degree of complexity as the Phaeophyceae. Very obvious protoplasmic connexions can be distinguished between the cells of nearly all forms except in the small group known as the Proto-florideae (cf. p. 217). Sexual reproduction is oogamous, the ovum being retained upon the parent plant, and although the subsequent development of the zygote is varied to a certain extent, it usually gives rise to filaments which bear special reproductive bodies or carpospores, and these latter are responsible for the production of the tetrasporic diploid plant. Most of the members exhibit a regular alternation of generations.

(10) NEMATOPHYCEAE. This is a fossil group of which one genus has been known for a long time (*Nematophyton*) whilst the other has only recently been described (*Nematothallus*). There is still considerable doubt as to their true affinities, but it would seem that

a place can best be found for them as a very highly developed type of alga. Their internal morphology would ally them closely with the more advanced members of either the Chlorophyceae or the Phaeophyceae. The only reproduction so far recorded is that of spores which were developed in tetrads, and therefore may have been akin to the Rhodophycean or Phaeophycean tetraspores.

REFERENCE

CARTER, P. W., HEILBRON, I. M. and LYTHGOE, B. (1939). *Proc. Roy. Soc.* B, **128**, 82.

CHAPTER II

CYANOPHYCEAE

*INTRODUCTION

This order used to be known as the Myxophyceae, but as this name was originally applied to a very heterogeneous group of organisms it is now customary to employ the name Cyanophyceae. The members of the group are characterized by a bluish green colour which varies greatly in shade, depending upon the relative proportions of chlorophyll a, β-carotin, myxoxanthin, phycocyanin and sometimes xanthophyll and phycoerythrin. The internal structure of the cell is extremely simple because a true nucleus and chromatophores are absent. Some authors have reported the presence of a nucleus with rudimentary chromosomes which undergo a form of mitosis, but these structures cannot be regarded as clearly established. The protoplast possesses two regions, a peripheral one containing the pigment together with oil drops and glycogen, and a colourless central area which contains granules. Two kinds of inclusions have been recognized. The metachromatic or α granules that lie in the colourless central area and which are nucleoproteic in nature since they give a Feulgen reaction. These granules have probably been mistaken by some workers for chromosomes, especially since it is found that they can divide by simple fission, although some authorities do not consider that this is even a primitive form of mitosis. The material of the central area is regarded by such workers as equivalent to the cytoplasm in the cells of higher plants. The other type of granule is known as the cyanophycin or β granule and occurs in the peripheral region. They are in the nature of a protein reserve, and their presence is probably dependent to a considerable extent upon the external environment.

The protoplast is normally devoid of vacuoles, and this fact may explain the great resistance of the plants to desiccation and of the cells to plasmolysis. In some forms, principally species which are planktonic, pseudo-vacuoles may be found, and it is supposed that these contribute towards their buoyancy. The protoplast is surrounded by an inner investment which has been shown to be a

modified plasmatic membrane. In addition there is an outer cell
sheath which may surround the whole cell, e.g. *Chroococcus*, or
form a cylindrical sheath, e.g. *Oscillatoria*, or an interrupted sheath,
e.g. *Anabaena*. This is usually composed of a pectic material,
although in the Scytonemataceae it may be made of cellulose. There
is considerable variation in the composition of the different cell
sheaths, and the amount of material laid down frequently depends
upon the external environment. In any case the secretion of pectins
by these plants is regarded as a primitive characteristic. In the
unicellular forms this material is produced at the periphery of the
cell, whilst in a few, e.g. *Chroococcus turgidus*, it accumulates in the
cytoplasm. Protoplasmic connexions between mature cells have
been recorded for one genus, *Stigonema*.

The group is characterized by a general absence of well-marked
reproductive organs; there are no sexual organs and no motile
reproductive bodies have ever been observed. It has recently
been suggested that the lack of sexuality can be correlated with the
absence of sterols in the group, an hypothesis that might well repay
further study. The coccoid forms (spherical cells) multiply by cell
division which takes place by means of a progressive constriction,
whilst in other types the cell contents divide up to give a number of
non-motile bodies that are termed *gonidia* (fig. 5). Crow (1922)
has pointed out that all stages from simple binary fission to gonidia
can be found:

(a) Binary fission, e.g. *Chroococcus turgidus*.

(b) Quadrants and octants formed, e.g. *C. varius*.

(c) Numerous small daughter cells are produced in which there
is a retention of individual sheaths, e.g. *Gloeocapsa* sp. and variants
of *Chroococcus macrococcus*.

(d) The same without individual sheaths, e.g. *C. macrococcus*,
Gloeocapsa crepidinum.

(e) Abstricted gonidia, e.g. *Chamaesiphon*.

Many of the filamentous forms produce specialized cells known
as *heterocysts*. These are enlarged cells which possess thickened
walls, and they usually occur singly though occasionally they may
be formed in rows. They develop from an ordinary vegetative cell,
but during development they remain in protoplasmic communication with neighbouring cells and if they contain contents, as they

probably do, these may be expected to differ from those of an ordinary vegetative cell. Various suggestions have been made as to their function, and in many cases they probably determine the breaking up of the *trichomes* (or threads) into *hormogones*. These hormogones are short lengths of thread which are cut off, thus forming a means of vegetative reproduction among the filamentous types. The heterocysts may also perhaps act as a food store, or they may represent archaic reproductive organs which are now function-less. It has been reported that in *Nostoc* and *Anabaena* these cells may occasionally behave as reproductive bodies. Hormogones, besides being cut off by the heterocysts, may also be produced by the development of biconcave separation disks which develop at intervals along the filament. The hormogones, together with certain of the filamentous types, exhibit a slow motion, and although cilia have been described for one species their presence has never been corroborated. The active and continual secretion of mucilage along the sides of the filaments is now regarded as the probable mechan-ism for securing movement. Thick-walled resting spores, or *akinetes*, occur in many of the filamentous forms, normally de-veloping next to a heterocyst. The entire lack of sexuality must be ascribed to the ancient cell structure and the absence of chromo-somes together, possibly, with the lack of sterols.

This type of cell structure naturally provides a problem for the geneticist. There are two possibilities because each cell may con-tain one single gene or a number of genes (organized self-reprodu-cing bodies which determine the properties of the cell and of the organism). The genes must be separated from each other since there are no chromosomes in which they could be situated, and they will either be distributed generally throughout the cell or else in a particular part of it. Since there is no special means of accurate partition sexuality would be useless because it could not confer the property of recombination but only of addition.

Many of the forms aggregate into colonies, but in some of the Chroococcaceae the plant mass is an association of such colonies and not one large colony or thallus. The form which any colony may take up depends on (1) planes of cell division, (2) effect of environment which may determine the consistency of the mucilage, uneven temperatures, for example, sometimes producing irregular growth. It has been shown experimentally that the environment may affect the shape of colonies of *Microcystis* and *Chroococcus turgidus* and determines the size of *Rivularia haematites* (cf. p. 337) Certain lines of morphological development have been followed by the group and the various lines may be depicted schematically as follows (cf. also fig. 1):

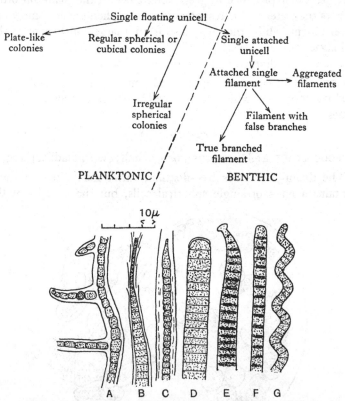

Fig. 1. Types of trichome in the Cyanophyceae. A, *Hapalosiphon arboreus*. B, *Calothrix parietaria*. C, *Schizothrix purpurascens*. D, *Oscillatoria margaritifera*. E, *O. proboscidea*. F, *O. irrigua*. G, *Arthrospira Jenneri*. (After Crow.)

As may perhaps be expected from a primitive group there is evidence of homoplastic or parallel development when compared with plants from other primitive groups, especially the Chlorophyceae. Homoplasy can be seen in *Gloeothece* and *Gloeocystis*, *Merismopedia elegans* and *Prasiola* (figs. 4, 40), *Chamaesiphon* and

Characium (figs. 5, 26), *Chroococcus* and *Pleurococcus* (figs. 3, 44), *Lyngbya* and *Hormidium*.

As a group, the plants are extremely widely distributed over the face of the earth under all sorts of conditions, frequently occurring in places where no other vegetation can exist, e.g. hot thermal springs. Their presence in great abundance in the plankton often colours the water and is responsible for the phenomenon known as water bloom, whilst they may also form a large constituent of the soil algae (cf. Chapter x).

The class is divided into two orders:

COCCOGONALES, which reproduce by means of single cells.

HORMOGONALES, which reproduce by groups of cells or hormogones.

COCCOGONALES

CHROOCOCCACEAE: *Microcystis* (*micro*, small; *cystis*, bladder). Fig. 2.

The thallus, which is free-floating, varies much in shape and contains a mass of single spherical cells, but the sheaths of the

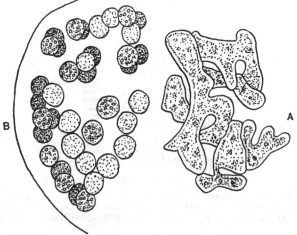

Fig. 2. *Microcystis aeruginosa.* A, colony. B, portion of a colony (× 750). (A, after Geitler; B, after Tilden.)

individual cells are confluent with the colonial envelope. Reproduction of the single cells takes place by means of fission in three planes, whilst reproduction of the colony is through successive disintegration, each portion growing into a new colony. The shape of the colony is primarily determined by the environmental conditions, and it can be changed by altering the environment artificially. *M. aeruginosa* is a very common water bloom alga.

*CHROOCOCCACEAE: *Chroococcus* (*chroo*, colour; *coccus*, berry). Fig. 3.

The cells are single or else united into spherical or flattened colonies each containing a small number of cells, the individual

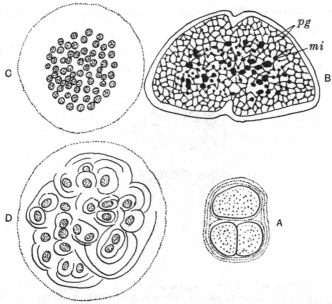

Fig. 3. *Chroococcus.* A, *C. turgidus,* plant (× 600). B, *C. turgidus,* protoplasmic reticulum with accumulations of metachromatin at nodal points. *pg* = plasmatic granules, *mi* = microsomes. C, *C. macrococcus,* normal daughter cell formation. D, *C. macrococcus,* daughter cell formation with retention of the parent envelopes. (A, after Smith; B, after Acton; C, D, after Crow.)

sheaths being homogeneous or, more frequently, lamellated. Plants grown in water produce a concentric envelope but when grown on damp soil the sheath is often asymmetrical. The outer integument is not very gelatinous and indeed is quite thin in some species. The colonies are either free-floating or else they form a layer on the soil. A study of the cytology of this genus has shown that *C. turgidus* represents the simplest condition with the metachromatin granules only just differentiated. In *C. macrococcus*, a more complex type, there is a central body which, according to Acton (1914), contains a fine reticulum with chromatin at the nodal points, but a reinvestigation of this species is perhaps desirable and might well lead to a different interpretation (cf. fig. 3). At cell division this "nucleus" divides by simple constriction, but there is no evidence of a mitosis. In *Gloeocapsa* a similar condition is observed, but in this case with evidence of a rudimentary mitosis.

CHROOCOCCACEAE: *Merismopedia* (*merismo*, division; *pedia*, plain).
    Fig. 4.

The free-floating colonies form regular plates one cell in thickness at first, but with increasing age they become irregularly square or

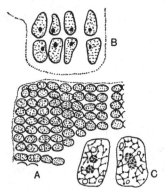

Fig. 4. *Merismopedia elegans.* A, portion of colony (× 345). B, portion of colony (× 1125). C, structure in cells about to divide (× 1875). (A, after Geitler; B, C, after Acton.)

rectangular and are often curved or twisted. The cells are spherical or ellipsoidal and their individual sheaths are confluent with the colonial envelope. There is every transition from compact (*M. aeruginosa*) to extremely loose colonies (*M. icthyolabe*), the number of cells enclosed in one envelope depending on the rate of division which only takes place in two planes. In *M. elegans*, prior to cell division, an accumulation of chromatin occurs in the centre of the cells to form a central body or so-called "nucleus" which divides by constriction immediately preceding cell division. The "nucleus" then disappears until the next division.

CHAMAESIPHONACEAE: *Chamaesiphon* (*chamae*, on the earth; *siphon*, a small tube). Fig. 5.

The cells are epiphytic, solitary, or arranged in dense clusters; they stand erect, are more or less rigid, vary much in shape and are attached at the base. The sheath is thin, hyaline and ultimately opens at the apex. Reproduction is by means of gonidia which are abstricted successively by transverse division from the apex, and as these gonidia have been regarded as one-celled hormogones the genus thus forms a link between the Coccogonales and Hormogonales.

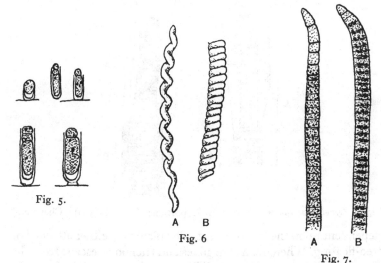

Fig. 5.

A B
Fig. 6

A B
Fig. 7.

Fig. 5. *Chamaesiphon cylindricus* with gonidia (× 1200). (After Geitler.)
Fig. 6. *Spirulina*. A, *S. major* (× 1070). B, *S. subsalsa* (× 1070). (After Carter.)
Fig. 7. *Oscillatoria*. A, *O. formosa* (× 613). B, *O. corallinae* (× 613). (After Carter.)

HORMOGONALES (REPRODUCTION BY HORMOGONES)

OSCILLATORIACEAE: *Spirulina* (*spirula*, a small coil). Fig. 6.

The trichomes have no proper sheath and are septate, although the septa are frequently very obscure. The trichomes are simple, free and coiled into a more or less characteristic spiral.

*OSCILLATORIACEAE: *Oscillatoria* (*oscillare*, to swing). Figs. 1, 7.

The trichomes are free, smooth or constricted, straight or arcuate and often form tangled masses, the sheaths being delicate or more frequently absent. The apical cell is sometimes provided with a cap or calyptra. There are a number of common species, *O. limosa* being frequently found on very damp soils, wet stones and other moist places.

Oscillatoriaceae: *Lyngbya* (after H. C. Lyngbye). Fig. 8.

This genus differs from *Oscillatoria* in the presence of a sheath of variable thickness and colour, the character of which is largely

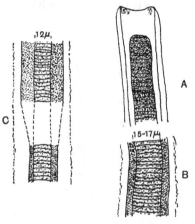

Fig. 8. *Lyngbya aestuarii.*   A, apex. B, C, portions of threads. (After Chapman.)

dependent upon the environment. The plants are either attached or free-floating. When the hormogones and trichomes escape from the sheaths it is frequently very difficult to determine whether they belong to *Oscillatoria, Lyngbya,* or some other similar genus.

Scytonemataceae: *Scytonema* (*scyto,* leather; *nema,* thread). Fig. 9.

The threads differ from those of the preceding genus in the presence of heterocysts. The filaments (trichome and sheath) have a base and apex, and the false branches arise either between two heterocysts or else adjoining a heterocyst. The intercalary growth results in strong pressure being applied to the sheath, which finally ruptures so that the trichome forms a loop outside (fig. 9 A–C). Further growth causes this loop to break, thus producing twin branches, one or both of which may subsequently proceed to additional growth, the branch sheaths extending back into the

parent sheath (fig. 9 E). More commonly, false branching is
initiated by degeneration of a vegetative cell or heterocyst and subsequent growth of the two filaments on either side.

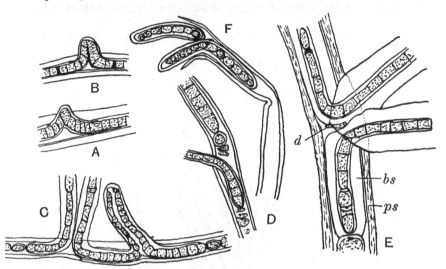

Fig. 9. Cyanophyceae. A–C, geminate branching in *Scytonema pseudoguyanense*
(A, × 470, B, C, × 340). D, false branching in *Calothrix ramosa* (× 570). E, false
branching in *Scytonema pseudoguyanense* showing branch sheath (*bs*) terminating
at heterocyst. *ps* = parent sheath, *d* = dead cell (× 590). F, hormogones emerging
from parent sheath in *S. guyanense* (× 750). (After Bharadwaja.)

*Rivulariaceae: *Rivularia* (*rivulus*, small brook). Fig. 10.

The colonies form spherical, hemispherical, or irregular gelatinous masses that are attached to plants or stones, those of *R. atra*
being especially frequent on salt marshes. They contain numerous
radiating filaments with repeated false branching, each branch
terminating in a colourless hair. The individual sheaths can be seen
near the base of the trichomes, but they are diffluent farther up.
The heterocysts are basal, and in one section of the genus spores are
produced next to them. The genus is also interesting because it
has been shown to contain xanthophyll.

*Nostocaceae: *Nostoc* (used by Paracelsus). Fig. 11.

The gelatinous thallus is solid or hollow, floating or attached,
and varies much in size and shape. There is a dense limiting layer

containing numerous intertwined and contorted filaments with individual hyaline or coloured sheaths which may be absent, indistinct or conspicuous. The heterocysts are terminal or inter-calary and are arranged singly or in series. Reproduction is by

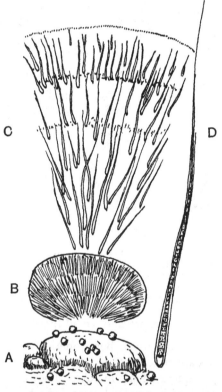

Fig. 10. *Rivularia atra*. A, plants on stones (× ⅔). B, transverse section of thallus (× 9). C, transverse section of thallus (× 45). D, single trichome in sheath (× 300). (A–C, after Newton; D, original.)

means of hormogones or spores, the latter arising midway between the heterocysts and developing centrifugally. *N. commune* forms gelatinous masses and is fairly common on damp soils.

The closely related genus *Anabaena* only differs from *Nostoc* in that no firm colony is formed. Some species are often symbiotic (cf. p. 297), whilst both *Anabaena* and *Nostoc* are apparently capable of fixing nitrogen from the atmosphere (cf. p. 304).

NOSTOCACEAE: *Cylindrospermum* (*cylindro*, cylinder; *spermum*, seed). Fig. 12.

A characteristic feature of this genus is the large spore which develops next to the heterocyst at one or both ends of a filament. The outer wall of the spore is often papillate.

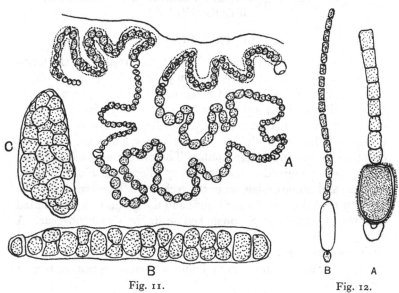

Fig. 11. Fig. 12.

Fig. 11. *Nostoc.* A, portion of colony of *N. Linckia* (×400). B, C, germinating hormogones of *N. punctiforme* (×900). (After Geitler.)

Fig. 12. *Cylindrospermum.* A, *C. majus* (×680). B, *C. stagnale* (×340). (After Geitler.)

REFERENCES

Chroococcus. ACTON, E. (1914). *Ann. Bot., Lond.*, **28**, 433.

Scytonema. BHARADWAJA, Y. (1933). *Arch. Protistenk.* **81**, 243.

Chroococcus, Gloeocapsa, Microcystis. CROW, W. B. (1922). *New Phytol.* **21**, 81.

General. CROW, W. B. (1924). *J. Genet.* **14**, 397.

General. CROW, W. B. (1928). *Arch. Protistenk.* **61**, 379.

Systematic. GEITLER, L. (1932). "Cyanophyceae" in Rabenhorst's *Kryptogamen Flora*, **14**, Leipzig.

General. POLJANSKI, G. and PETRUSCHEWSKY, G. (1929). *Arch. Protistenk.* **67**, 11.

Cytology. SPEARING, J. K. (1937). *Arch. Protistenk.* **89**, 209.

CHAPTER III

CHLOROPHYCEAE

VOLVOCALES, CHLOROCOCCALES, ULOTRICHALES, OEDOGONIALES

*INTRODUCTION

The older botanists included in the term Chlorophyceae the forms
which are now placed in the Xanthophyceae (cf. p. 113), but in
1897 Bohlin pointed out that some of the green algae possessed
unequal cilia, and in 1899 Luther coined the term "Heterokontae"
for such forms. In 1902 Blackman and Tansley revised the classification of the green algae using the terms Isokontae, Akontae,
Stephanokontae and Heterokontae. These were adopted by most
workers and remained in use until 1927 when Fritsch included the
Akontae and Stephanokontae in the Isokontae. The term Isokontae
has thus ceased to be of significance and the group is now included
with the Akontae and Stephanokontae in the Chlorophyceae. A
division into two great groups, marine and fresh water, as suggested
by Tilden in 1935, is not at all feasible, because nearly all the
morphologically distinguishable families possess representatives in
both environments.

The cell structure is fairly characteristic, the protoplast often
containing a large central vacuole, which in the simpler forms is
contractile and serves to remove surplus water and waste matter.
The green pigment, which is essentially identical with that of the
higher plants, is contained in plastids: there is usually only one of
these in a cell and its outline may be discoid, star-shaped, spiral,
plate-like or reticulate. There is some evidence to show that these
plastids are capable of movement in response to light stimuli.
Other colouring matter may also be present, e.g. haematochrome in
Sphaerella and phycoporphyrin in some of the Zygnemales, whilst
fucoxanthin (cf. p. 129) is found in *Zygnema pectinatum*. The cells
are commonly surrounded by a two-layered wall, the inner, which is
often lamellate, being of cellulose, and the outer of pectin, but in
some forms the outer surface of this pectin sheath is dissolved as
fast as it is formed on the inner side. In a few species there is a third

layer or cuticle, whilst in others there is an outer mucilage layer, and in at least three groups (Siphonales, Siphonocladiales and Charales) lime may be deposited on the walls. The chloroplasts normally contain rounded bodies, or *pyrenoids*, which are composed of a viscous mass of protein. The pyrenoids are usually surrounded by a starch sheath, starch being the principal product of photosynthesis, and it is said that parts of the pyrenoid are successively cut off to form starch grains, but the evidence for this is not entirely satisfactory. The pyrenoids are perpetuated by simple division but they may also arise *de novo*.

Each cell usually contains one nucleus, but in certain groups a multinucleate condition is to be found. Each nucleus possesses a deeply staining body, the *nucleolus*, together with chromosomes which are usually short and few in number, although these latter may be masked during the interphases between nuclear division. At cell division the pyrenoids and chloroplasts may also undergo division. The flagellae of the motile bodies are composed of an axial cytoplasmic filament surrounded, except at the very apex, by a sheath which probably has the power of contraction, whilst in the Volvocales the flagellae normally disappear at the commencement of cell division. The motile cells also possess a red eye-spot, the detailed structure of which is not yet elucidated in all the groups, though it appears to contain a primitive lens in the Volvocales. The red colouring matter is due in part to the chromolipoid pigment known as haematochrome (cf. fig. 13).

Vegetative reproduction takes place through fragmentation and ordinary cell division, whilst asexual reproduction is by means of bi- or quadriflagellate zoospores which are commonly produced in normal cells because special sporangial structures are rare. These zoospores are often formed during the night and are then liberated in the morning: after liberation they may remain motile for as much as 3 days or for as short a time as 3 min. Their production can sometimes be artificially induced by altering the environmental conditions, e.g. removing the plant from flowing to still water (*Ulothrix, Oedogonium*), changing the illumination, transferring to water from air (terrestrial *Vaucheria* spp.), or removing from water for 24 hours (*Ulva, Enteromorpha*). Each individual cell may produce one or more zoospores, the number varying with the different species. Liberation is secured by means of (*a*) lateral pores, (*b*) terminal pores, (c) gelatinization of the entire wall, (d) the wall dividing into two equal or unequal halves. In some species non-motile zoospores are formed which are called *aplanospores*, but if these should then secrete a thick wall they become known as *hypnospores*. Aplanospores which have the same shape as the parent cell are termed *autospores*. All these spores develop a new membrane when they are formed and hence differ from a purely resting vegetative cell or *akinete* (cf. fig. 13).

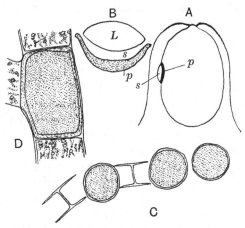

Fig. 13. A, diagram of eye-spot of *Chlamydomonas*. p=pigment cup, s=photosynthetic substance. B, diagram of cross-section of eye-spot of *Volvox*. L=lens, p=pigment cup, s=photosynthetic substance. C, aplanospores of *Microspora Willeana* (×600). D, akinete of *Pithophora oedogonia* (×225). (After Smith.)

Sexual reproduction is represented in all the orders and often there is a complete range from isogamy to oogamy, the ova usually being retained on the parent thallus in the oogamous forms (e.g. *Vaucheria, Coleochaete*). The isogamous forms are normally dioecious, the two strains being termed + and −, and as they are usually alike morphologically they can only be distinguished by the behaviour of the gametes. In some cases (*Ulva*) relative sexuality is known to occur, weak + or − strains fusing with strong + or − strains respectively. Indeed, Hartmann (1924) has declared that all gametes are potentially bisexual, and there would seem to be considerable grounds for supporting this view. Segregation into + and − strains occurs during meiosis, a phenomenon which in many species takes place at the first or second division of the zygote.

The occurrence of sexual reproduction in Nature often marks the phase of maximum abundance when the climax of vegetative activity has just been passed. It can also be brought on in culture by an abundance or deficiency of food material or by intense insolation. Interspecific hybrids have been recorded from *Spirogyra, Ulothrix, Stigeoclonium, Draparnaldia* and *Chlamydomonas* (fig. 14). Another striking fact is that characters which may develop in some species under the influence of the external environment are normally found "fixed" in others. This not only indicates the plasticity of many members in the class, but the phenomenon might also be of importance in considerations of phylogenetic relationships.

Vegetative evolution would appear to have taken place along several lines and may be represented schematically thus:

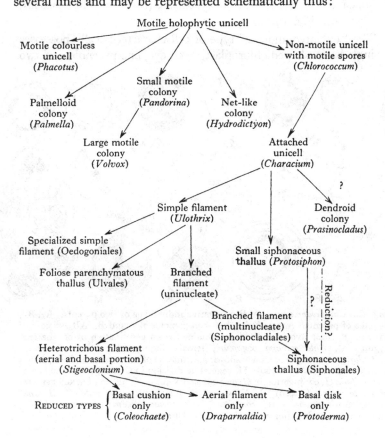

The names of the species and genera do not imply that they provided the actual links in the process of evolution, but that ancestral forms having an appearance similar to that of the examples quoted formed the intermediate stages. The examples are given in order that the student may have something concrete upon which to visualize the scheme.

As a group the Chlorophyceae are very widespread, occurring in all types of habitat. A few species, e.g. *Endoderma*, *Chlorochytrium*, *Rhodochytrium*, are parasitic, whilst several other species participate in symbiotic associations, e.g. *Carteria*, *Zooxanthella*, *Chlorococcum* (cf. p. 296).

VOLVOCALES

*CHLAMYDOMONADACEAE: *Chlamydomonas* (*chlamydo*, cloak; *monas*, single). Fig. 14.

The "chlamydomonad" type of cell characteristically possesses a single basin-shaped chloroplast, a red eye, one pyrenoid and two

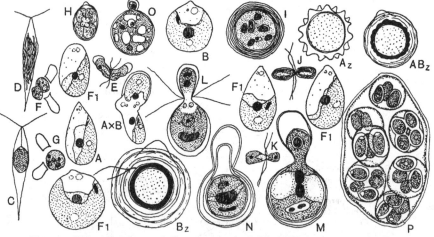

Fig. 14. *Chlamydomonas*. A, B, vegetative individuals of two parents. A_z, B_z, zygotes of parents. A × B, fusion between gametes of A and B. AB_z, zygote of hybrid. F_1, four hybrid individuals obtained from germination of one hetero-zygote. C, *Chlorogonium oogamum*, female showing formation of ovum. D, *Chlorogonium oogamum*, male showing formation of antherozoids. E–G, stages in fusion of *C. media* (× 400). H, vegetative division in *C. angulosa*. I, zygote of *C. coccifera*. J, conjugation in *C. longistigma* (× 400). K, fusion of naked gametes of *C. pisiformis* (× 400). L–N, stages in fusion of gametes of *C. Braunii*. O, fusion of gametes in *C. coccifera*. P, *C. Braunii*, palmelloid stage. (A–D, after Fritsch; E–K, after Scott; L–P, after Oltmanns.)

flagellae, and is often strongly phototactic. Variations in the
structure of the cell occur throughout the genus, which contains
about 150 species. There may be more than one pyrenoid present
(*C. sphagnicola*) or they may be completely absent (*Chloromonas*),
whilst the chloroplast may be reticulate (*Chlamydomonas reticulata*),
or axile and stellate (*C. eradians*), or it may be situated laterally
(*C. parietaria*). It has been said that under cultural conditions many
of the characteristic features can be modified, and that therefore
some of the forms are not true species but are simply phases in
the life cycles of other species.

The motile cells are spherical, ellipsoid, or pyriform in shape with
a thin wall which occasionally possesses an outer mucilage layer.
The two flagellae are situated anteriorly and either project through
one aperture in the wall or else through two separate canals, but in
either case at the point of origin of the flagellae there are two basal
granules whose function is not yet clearly established. Each cell
typically possesses two contractile vacuoles which have an excretory
function. At asexual reproduction the motile bodies come to rest
and divide up into four, more rarely eight or sixteen, daughter cells.
The first division at zoospore formation is normally transverse, and
in those cases where it is longitudinal a subsequent twisting of the
protoplast makes it appear to be transverse. The zoospores escape
through gelatinization of the cell wall, but if this does not occur the
colony then passes into the palmelloid state, which is usually of
brief duration, though in *C. Kleinii* it forms the dominant phase in
the life history of the species. *C. Kleinii* may thus be regarded as
forming a transition to the condition found in *Tetraspora* (cf. p. 34).

In sexual reproduction eight, sixteen or thirty-two gametes are
formed in each cell. In *Chlamydomonas longistigma* the gametes are
bare (*gymnogametes*); in *C. media* they are enclosed in a cell wall
from which they emerge in order to fuse (*calyptogametes*); in
C. monoica there is anisogamy as the naked contents of one gamete
pass into the envelope of the other; in *C. Braunii* there is marked
anisogamy, the female cell producing four macrogametes and the
male cell eight microgametes; in *C. coccifera* there is oogamy, with
the female cell producing one macrogamete enclosed in a wall whilst
the male cell produces sixteen spherical microgametes. In a related
genus, *Chlorogonium oogamum*, one naked ovum is produced and
numerous elongate antherozoids, whilst cases of relative sexuality

have been recorded for *Chlamydomonas eugametos*. The zygote on germination frequently gives rise to four swarmers, and it is probable that meiosis occurs during this segmentation, the normal vegetative cells thus being haploid. In *C. pertusa* and *C. botryoides*, however, the zygote may remain motile for as long as 10 days, and hence it may be considered that these two species exhibit a definite alternation of generations. In *C. variabilis* the persistent quadriflagellate zygote has for long been known as *Carteria ovata* (cf. also p. 297), but it has now been shown that the latter is the diploid generation of the *Chlamydomonas*.

The genus is widespread, the various species occurring principally in small bodies of water or on the soil.

CHLAMYDOMONADACEAE: *Gonium* (*gonium*, angle). Fig. 15.

The colony in the different species is composed of four, eight or sixteen cells all lying in one plane and forming a flat quadrangular plate, but it has been suggested that the four- and eight-celled

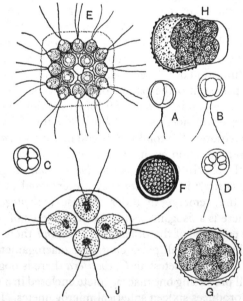

Fig. 15. *Gonium pectorale.* A–D, stages in the formation of a coenobium. E, colony (× 520). F, zygote. G, H, stages in germination of zygote. J, four-celled colony. (A–D, after Fritsch; E, after Smith; G–J, after Kniep.)

colonies are merely degenerate forms of the principal species, *G. pectorale*. In the sixteen-celled colonies (*G. pectorale*) there are four cells in the centre and twelve in the periphery, each cell being surrounded by a gelatinous wall and fused to the neighbouring cells by means of protrusions, whilst the protoplasts of the individual cells are also united by fine protoplasmic threads. The ovoid or pyriform cells contain contractile vacuoles and are provided with a pair of flagellae. The centre of the colony is composed of mucus and there is also a firm outer gelatinous layer. The shape of the colony accounts for its mode of progression which is by means of a series of somersaults around the horizontal axis. At asexual reproduction all the cells divide simultaneously to form daughter colonies. If single cells should become isolated then after a time they will give rise to (*a*) daughter colonies, (*b*) akinetes, or (*c*) a palmelloid state. Sexual reproduction is by means of naked isogametes, fusion occurring between gametes from separate colonies as the various species occur in + and − strains. The resulting quadriflagellate zygote soon comes to rest and subsequently germinates, when it gives rise to four biflagellate haploid cells which are liberated together as a small colony. When the later development of these cells is followed it is found that two of them give rise to + and two to − colonies, suggesting that meiosis must take place at germination of the zygote.

*CHLAMYDOMONADACEAE: *Pandorina* (after Pandora's box). Fig. 16.

The colonies are oblong or spherical and are composed of four, eight, sixteen or thirty-two cells, sixteen being the normal number in the common species *P. morum*. The cells, which are arranged

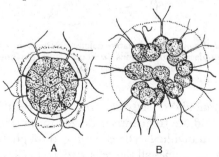

A   B

Fig. 16. *Pandorina morum*. A, vegetative colony (× 975). B, colony with female gametes (× 975). (After Smith.)

compactly in the centre and are frequently flattened from mutual pressure, are connected to each other by protoplasmic threads that are withdrawn during reproduction. Each colony is enclosed in a gelatinous matrix with an outer watery sheath, and, together with the next two genera, exhibits some degree of polarity in its progression. When reproducing asexually the cells first lose their flagellae and then each one gives rise by several divisions to a daughter colony. In sexual reproduction signs of anisogamy are to be found, and the zygote germinates giving one to three biflagellate spores which then develop into new colonies.

*CHLAMYDOMONADACEAE: *Eudorina* (*eu*, well; *dorina*, meaningless!).
Fig. 17.

The colonies are spherical or ellipsoid, the posterior end often being marked by mamillate projections. They contain sixteen,

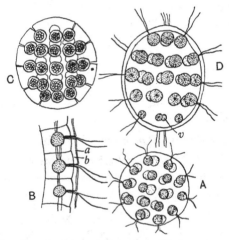

Fig. 17. *Eudorina elegans*. A, vegetative colony. B, transverse section showing structure and protoplasmic connections. *a* = outer layer, *b* = inner layer of mucilage. C, formation of daughter coenobia. D, *E. illinoiensis*, showing somatic cells, *v*. (After Fritsch.).

thirty-two (commonly) or sixty-four biflagellate cells, which are not closely packed and are frequently arranged in transverse rows, the flagellae of the individual cells emerging through funnel-shaped apertures. In most species all the cells give rise to daughter colonies, but in *E. illinoiensis* and *E. indica* the four anterior cells are much smaller and cannot produce gametes or daughter colonies. This marks a first differentiation into a plant soma within the group, and furthermore these somatic cells die once the colony has reproduced. It would be of great importance if the nature of the stimulus that induced some of the cells to lose their reproductive capacity could be determined. It might be possible to investigate such a problem experimentally on some of the undifferentiated species of *Eudorina*. Sexual reproduction is oogamous, the colonies being either monoecious or dioecious: in the former case the anterior cells give rise to the antherozoids, whilst in the latter case the antheridial plates are liberated intact and only break up after swimming to the female colony where the surrounding walls have already become gelatinous. The zygote on germination gives rise to one motile zoospore and two or three degenerate zoospores.[1]

CHLAMYDOMONADACEAE: *Pleodorina* (*pleo*, more; *dorina*, meaningless!). Fig. 18.

This genus is very similar to the preceding one, but the somatic area is more highly differentiated as it occupies one-third to one-

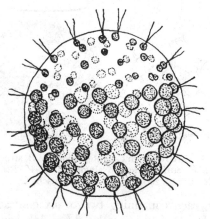

Fig. 18. *Pleodorina Californica.* Colony of 120 cells (× 178). (After Shaw.)

half of the colony, and the total number of cells is greater, thirty-two, sixty-four or 128. The somatic cells are all situated either in an anterior or posterior position and they die when the colony has reproduced. Reproduction follows the same lines as in *Eudorina*.

[1] Inversion of the daughter colonies and of the antheridia takes place during development (cf. *Volvox*).

*CHLAMYDOMONADACEAE: *Volvox* (*volvere*, to roll). Figs. 19–22.

This genus represents the ultimate development that has been reached along this particular line, each colony forming a hollow sphere with 500–20,000 biflagellate cells set around the periphery, the flagellae emerging through canals. The interior of the colony is mucilaginous or else merely contains water, whilst the whole collection of cells is bounded by a firm mucilage wall. The

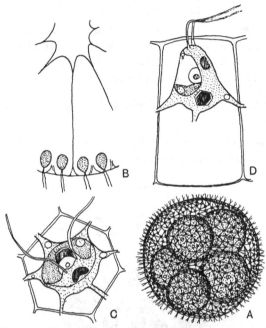

Fig. 19. *Volvox*. A, *V. aureus* with daughter colonies. B, structure of *V. aureus* as seen in section. C, surface view of single cell of *V. Rousseletii* (× 2000). D, the same in side view (× 2000). (A–B, after Fritsch; C, D, after Pocock.)

individual cells, each containing two to six contractile vacuoles, are surrounded by gelatinous sheaths, the middle lamellae of which form a polygonal pattern when stained with methylene blue. The cells are usually united by two or more delicate cytoplasmic threads, or *plasmodesmae*, though these are absent in some species (*V. tertius*). In *V. globator* the cells are sphaerelloid in nature, whilst in *V. aureus* they are chlamydomonad in appearance, several individual chloroplasts being enclosed in wedge-shaped prisms which are

probably morphologically equivalent to cells. For this reason it has
been suggested that the volvocine colony has arisen at least twice in

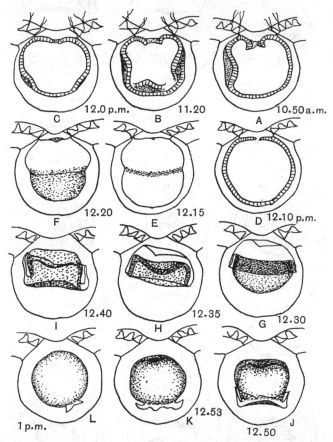

Fig. 20. *Volvox capensis* and *V. Rousseletii*. A–J, stages in the inversion of a
daughter colony. A, denting begins. C, dents smooth out. D, colony round
again. E, 'hour-glass' stage. F, posterior half contracts. G, infolding begins.
H, infolding complete. I, posterior half emerges through phialopore. J, flask
stage begins. K, flask stage ends. L, inversion complete. (All × 150 approx.)
(After Pocock.)

the course of evolution, once from a *Sphaerella* and once from a
Chlamydomonas ancestry. On the other hand, the great uniformity
of their sexual reproduction can be employed as an argument
against such a diphyletic origin.

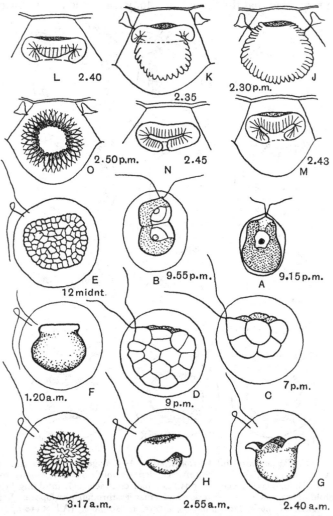

Fig. 21. *Volvox.* A–I, stages in the development of the zoospore of *V. Rousseletii.* A, zoospore just after escape. B, first division. F, preparation for inversion. G–I, inversion. (All × 375.) J–O, stages in the inversion of a sperm bundle of *V. capensis.* (All × 750.) (After Pocock.)

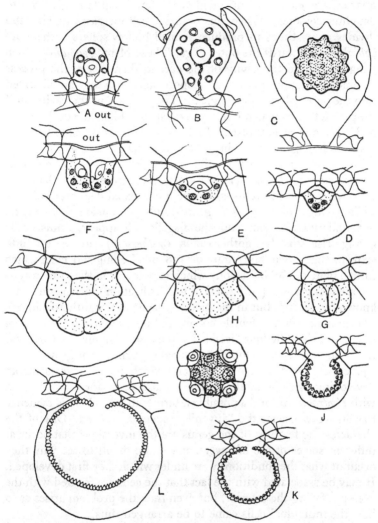

Fig. 22. *Volvox*. A–C, development of oospore of *V. Rousseletii* (× 750). *out* = outer wall. A, flagellar stage. B, mature. C, exospore formation. D–L, development of daughter colony (gonidium). F, two-celled stage (× 750). G, four-celled stage (× 750). H, eight-celled stage (× 750). I, sixteen-celled stage (× 750). J–L, formation of phialopore (× 225). (After Pocock.)

The majority of the cells, including *all* those in the anterior quarter, are wholly somatic, and only a few are able to give rise to daughter colonies. When this occurs a cell increases in size and divides many times to produce a small hollow sphere with a pore (*phialopore*) towards the outer edge. These plants (*gonidia*), which hang down into the cavity of the parent, then invert, the process commencing opposite their phialopore, and later they are liberated into the parental cavity (cf. fig. 20). They remain in the cavity until the parent tears open, in *Volvox aureus* at the phialopore of the adult, in *V. globator* at any place. In *V. africana* it is possible to see as many as four generations in the one original parent colony.

In sexual reproduction the plants are either monoecious (*V. aureus*) or dioecious (*V. globator*), and, furthermore, plants reproducing sexually are usually devoid of asexual daughter spheres. Cells giving rise to eggs (*egg cells*) enlarge considerably, but do not undergo division, and the flagellae disappear, whilst cells giving rise to the antherozoids (*antheridia*) divide up into sixteen, thirty-two, sixty-four or 128 small elongated cells which form a plate or globoid colony which may invert in the same way as the asexual gonidia (cf. fig. 21). The fertilization mechanism is not known for certain, but in the dioecious species the antherozoids are said to penetrate the female colony and then enter the ovum from the inner side. The first divisions of the zygote involve meiosis, and the oospore then develops into a single swarmer that grows into a "juvenile" plant of about 500 cells which finally inverts before developing into the adult (cf. fig. 21, also p. 43 for a comparison with *Hydrodictyon* and a possible interpretation). There is evidence that in some species the "juvenile" stage is omitted. One of the characteristic features of the genus are the inversions that occur at different stages of the life cycle, and it is difficult to see why they occur or what the conditions were under which they first developed. It may be associated with the fact that the cells are formed with the eye-spot facing the interior, but even then the problem arises as to how the individual cells came to be arranged thus.

*Sphaerellaceae: *Sphaerella* (*sphaer*, ball; *ella*, diminutive of affection) (*Haematococcus*). Fig. 23.

A characteristic of this genus is the area between the protoplast and the cell wall; this is filled by a watery jelly and is traversed by cytoplasmic threads passing from the central protoplast to the cell wall. The protoplast contains several contractile vacuoles and one or more pyrenoids, although two is the usual number. Asexual reproduction is by means of two to four macrozoospores, whilst sexual reproduction is isogamous or anisogamous. The young *Sphaerella* cell is very akin to a *Chlamydomonas*, and for this reason some authors would unite the two genera. Large akinetes are

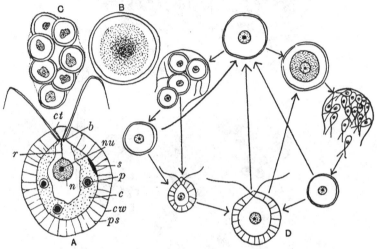

Fig. 23. *Sphaerella lacustris* (*Haematococcus pluvialis*). A, diagram of single macrozoid. *b* = blepharoplast, *c* = chloroplast, *ct* = flagellum tube, *cw* = cell wall, *n* = nucleus, *nu* = nucleolus, *p* = pyrenoid, *ps* = protoplasmic strand, *r* = rhizoplast, *s* = stigma. B, encysted plant with haematochrome in centre. C, eight-celled palmelloid stage. D, diagram illustrating life cycle in bacteria-free cultures. (After Elliott.)

known which on germination give rise to zoospores, hypnospores, or gametes. One species forms one of the components of "Red Snow" because under nival conditions it develops haematochrome as a result of nitrogen deficiency brought about by the presence of the snow. Periodic drying also appears to be an essential factor if the life history of the common species, *Sphaerella lacustris*, is to be maintained. Eight-celled colonies (*coenobia*), which behave just like *Pandorina*, are known in the related genus *Stephanosphaera*.

*TETRASPORACEAE: *Tetraspora* (*tetra*, four; *spora*, spores). Fig. 24.

The members of this genus form expanded or tubular, convoluted, light green macroscopic colonies. These are most abundant in the spring when they are attached at first, although later they become free-floating. The cells are embedded in the mucilage in groups of four, each group often being enclosed in a separate envelope. Two or four pseudocilia proceed from each cell to the surface of the main colonial envelope, each thread being surrounded by a sheath of

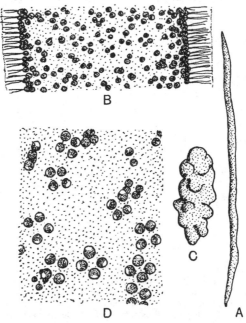

Fig. 24. *Tetraspora.* A, *T. cylindrica* (× ½). B, portion of colony of *T. cylindrica* showing outer envelope (× 155). C, *T. lubrica* (× ½). D, portion of colony of *T. lubrica* (× 500). (After Smith.)

denser mucilage. These structures cannot be organs of locomotion because they possess no power of movement, but they may represent such organs which have lost their function or they may be their precursors. Reproduction is either by fragmentation of the parent colony or else by means of biflagellate swarmers which may develop into (*a*) a new colony, (*b*) the palmelloid state or (*c*) a thick-walled resting spore. The resting spore gives rise to an amoeboid cell on germination. Sexual reproduction is secured by means of biflagellate isogametes, the colonies being either monoecious or dioecious, and after fusion has taken place the zygote divides into four to eight aplanospores which later grow into new colonies. The place of meiosis in the life cycle is not yet known.

CHLORODENDRACEAE: *Prasinocladus* (*prasino*, leek-like; *cladus*, shoot) (*Chlorodendron*). Fig. 25.

This genus is to be found principally in marine aquaria where it starts life as a quadriflagellate swarmer of the chlamydomonad

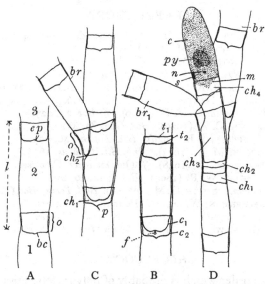

Fig. 25. *Prasinocladus*. A, B, portion of plant showing cell structure. 1–3 = cells. C, portion of plant showing arrangement of cells at branching. D, portion of plant with branches and living cell. (All × 1600.) py = pyrenoid, n = nucleus, s = stigma, c = chloroplast, m = basal margin of terminal protoplast, br = first branch, br_1 = second branch, ch_1–ch_4 = short chambers behind terminal cell at times of division, ch_1 being the earliest, f = minute remnant of flagellae, c_1, c_2 = bases of two cells, p = papilla, o = overlap of lateral wall, l = entire lateral extent of one chamber, ep = papilla pointing upwards, bc = basal cross wall, t_1, t_2 = tops of two cells. (After Lambert.)

type. The swarmer comes to rest and a new wall is formed with papillae at the base. Then the apex of the old wall ruptures, and when the contents have developed flagellae they move up, together with the new wall, so that the new cell becomes enclosed in the neck of the old one. The flagellae are lost for a time and then the process is repeated, and in this manner a filament of dead cells is built up

with a living cell at the apex. An oblique division of the living cell
results in a branch being formed and sometimes one half may cease
to divide, thus leaving a living cell in the middle of the dead cells.
It is evident from a consideration of this process that at each division a potential swarmer is formed which is not normally liberated.
On the few occasions when it is freed then the species is perpetuated, but at present the particular conditions under which a swarmer
may be liberated are not known.

REFERENCES

Eudorina. AKEHURST, S. C. (1934). *J. Roy. Micr. Soc.* **54**, 99.
Cytology. CHAUDEFAUD, M. (1936). *Rev. Alg.* **8**, 5.
Sphaerella. ELLIOTT, A. M. (1934). *Arch. Protistenk.* **82**, 250.
Tetraspora. GEITLER, L. (1931). *Biol. Zbl.* **51**, 173.
Gonium. HARPER, R. A. (1912). *Trans. Amer. Micr. Soc.* **31**, 65.
Eudorina, Gonium. HARTMANN, M. (1924). *Arch. Protistenk.* **49**, 375.
Prasinocladus. LAMBERT, F. D. (1930). *Z. Bot.* **23**, 227.
Volvox. LANDER, C. A. (1929). *Bot. Gaz.* **87**, 431.
Flagellae. PETERSEN, J. B. (1929). *Bot. Tidsskr.* **40**, 373.
Volvox. POCOCK, M. A. (1933). *Ann. S. Afr. Mus.* **16**, 523.
Volvox. POCOCK, M. A. (1938). *J. Quekett Micr. Club*, ser. 4, **1**, 1.
Pleodorina. SHAW, W. R. (1894). *Bot. Gaz.* **19**, 279.
Eudorina, Gonium. SMITH, G. M. (1930–1). *Bull. Torrey Bot. Club*, **57**, 359.
Volvox. ZIMMERMAN, W. (1921). *Jb. wiss. Bot.* **60**, 256.
Chlamydomonas. BEHLAU, J. (1939). *Bei. Biol. Pflanzen*, **27**, 221.
Eudorina. DORAISWAMI, S. (1940). *Journ. Ind. Bot. Soc.* **19**, 113.

CHLOROCOCCALES

This is an order which is probably of polyphyletic origin but it is
not proposed to elaborate this problem here.

*CHLOROCOCCACEAE: *Characium* (a slip or cutting). Fig. 26.

Each plant is a solitary unicell and only possesses a motile reproductive phase, and it may be supposed that in some previous era
the vegetative phase ceased to be motile and became attached. The
ellipsoidal cells occur singly or in aggregates on submerged plants
or living aquatic larvae, being borne on a short stalk which emerges
from a small basal disk. Asexual reproduction is brought about by
means of biflagellate zoospores which are liberated through a
terminal or lateral aperture. Certain species exhibit anisogamy,
whilst in *C. saccatum* the sexual and asexual generations are distinct
so that there may therefore be two different cytological generations.

*CHLOROCOCCACEAE: *Chlorochytrium* (*chloro*, green; *chytrium*, vessel). Fig. 27.

The swarmers, which may either be zoospores or motile zygotes, settle on the leaves of aquatics, principally species of *Lemna* (*Chlorochytrium lemnae*), whilst another species is also known which penetrates the leaves of *Polygonum lapathifolium*. Tubular prolongations grow out from these attached bodies and enter the host, either by way of the stomata or else between two epidermal cells.

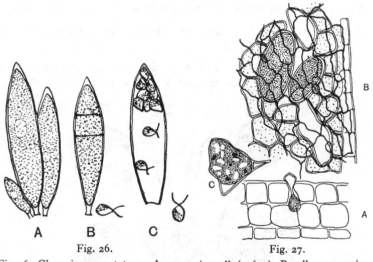

Fig. 26.   Fig. 27.

Fig. 26. *Characium angustatum*. A, vegetative cells (× 650). B, cell commencing zoospore formation (× 650). C, liberation of zoospores: the cell is probably broken accidentally (× 650). (After Smith.)

Fig. 27. *Chlorochytrium lemnae*. A, entrance of zygote into host. B, resting cells in leaf of *Lemna*. C, resting cell. (After Fritsch.)

Subsequently the end of the tube swells out into an ellipsoidal or lobed structure into which the contents of the swarmer pass. These swellings, which are to be found in the intercellular spaces of the host's tissues, become rounded off, and in the autumn sink down with the *Lemna* fronds to remain dormant until the next spring. In *Polygonum* the swollen filaments even crush the host cells which may become partially dissolved. In the spring the cell contents divide up into biflagellate swarmers, which are probably haploid, and these are liberated all together in a mucilaginous vesicle, fusion taking place whilst still enclosed or else after they have escaped. The resulting zygote is motile and quadriflagellate. Swarmers are also known which do not fuse, and it has been suggested that these develop from haploid races which have arisen apogamously, but a simpler explanation would be to regard them as zoospores. In any case the principal phase in the life cycle would seem to be diploid. Resting cells are also known in which the walls are thick and stratified. Species have been reported from mosses and algae as well as angiosperms, and as many of them have a decided pathogenic action they must at least be facultative parasites.

*CHLOROCOCCACEAE: *Chlorococcum* (*chloro*, green; *coccum*, berry) (*Cystococcus*). Fig. 28.

Much confusion has existed over this genus, as many of the species formerly described are now known to be phases in the life

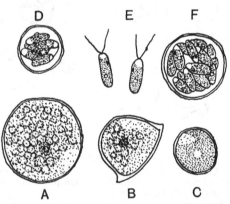

Fig. 28. *Chlorococcum humicolum.* A–F, various stages in the life history (× 800). (After Smith.)

cycles of species from other genera. Some of the species have been segregated into the genus *Trebouxia*, the cells of which form the algal component of several lichens (cf. p. 296). The plants are non-motile spherical cells which vary much in size, occurring singly or else forming a stratum on the soil. There is no eye-spot or con-tractile vacuole; the chloroplast is parietal, and there may be one or more pyrenoids. The cell walls are two-layered with a thin inner layer and an outer gelatinous one which is sometimes lamellose. The young cells are uninucleate but the adult ones are commonly multinucleate, and it is in this older condition that the protoplast divides and gives rise to numerous biflagellate zoospores which are liberated all together in a vesicle, usually in the early hours of the morning. After a short motile phase the flagellae are withdrawn and a new vegetative phase commences.

Isogamy and anisogamy are known, but there is no recorded example of even primitive oogamy comparable to that found in *Chlamydomonas*. Under certain conditions aplanospores are formed: when this happens the parent gelatinizes and a "palmella" stage results, the cells of which subsequently give rise to two to four biflagellate gametes. It seems clear that the suppression of motility has occurred several times in the Chlorococcaceae, a feature which supports the idea of their polyphyletic origin. The aplanospore stage also suggests how the genus *Chlorella* may have arisen. Under normal conditions *Chlorococcum* reproduces by means of motile zoospores, but when subjected to drought these bodies are non-motile. In nutrient culture solutions of low concentration reproduction takes place by zoospores, whilst in highly concentrated solutions the zoospores are replaced by aplanospores, so that it can be concluded that the environment may affect the reproductive mechanism to a considerable extent. *C. humicolum* is a very common soil form (cf. p. 299).

CHLORELLACEAE: *Chlorella* (*chlor*, green; *ella*, diminutive of affection). Fig. 29.

The globular cells are non-motile, solitary or aggregated into groups, and usually lack pyrenoids. They reproduce by division

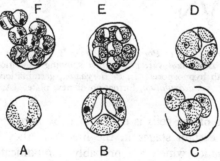

Fig. 29. *Chlorella vulgaris*. A, single cell. B, division into four. C, final stage of division into four daughter cells. D, first stage of division into eight. E, F, second and third stages of division into eight daughter cells. (After Grintzesco.)

into two, four, eight or sixteen autospores. Several species often form a symbiotic association with lower animals when they are known as *Zoochlorella* or *Zooxanthella* (cf. p. 296). The species are frequently indeterminate systematically and are chiefly studied by means of laboratory cultures, but in spite of these systematic difficulties they are common objects for physiological experiments.

HYDRODICTYACEAE: *Pediastrum* (*pedia*, plain; *astrum*, star). Fig. 30.

The species of this genus are common components of freshwater plankton. The cells form disk-like coenobia, the plane-faced or lobed cells being arranged in one layer, experiments suggesting

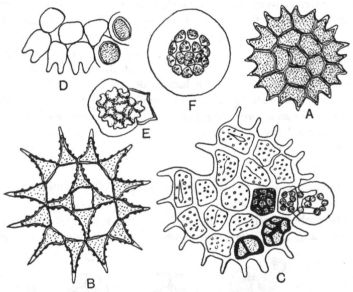

Fig. 30. *Pediastrum.* A, *P. Boryanum* (× 333). B, *P. simplex* var. *duodeniarum* (× 333). C, *P. Boryanum* var. *granulatum* showing liberation of zoospores. D, *P. duplex* with hypnospores. E, *P. Boryanum*, germination of tetrahedron. F, *P. Boryanum* var. *granulatum*, formation of new plate. (A, B, after Smith; C–F, after Fritsch.)

that the shape of the cells is determined by heredity and mutual pressure. At certain stages in the life cycle they bear tufts of gelatinous bristles which are probably a modification for their floating existence. There are 2–128 cells in each coenobium, varying with the species, and whereas the young cells are uninucleate the mature ones may possess as many as eight nuclei. Biflagellate zoospores are formed, the number depending upon the external physical conditions, and they are usually liberated at daybreak from the parent cell into an external vesicle in which they swarm for a time, but they soon become arranged into a new coenobium before the vesicle ruptures. The flagellae are sometimes absent. Isogametes are also formed and liberated singly, and after fusion the zygote divides up into a number of swarmers; each of these subsequently turns into a thick-walled polyhedral cell in which a new coenobium is formed. There would seem to be very little justification for placing this and the next genus into the Siphonales, as some authors have suggested, because their mode of reproduction is essentially much more akin to that of the Chlorococcales.

*HYDRODICTYACEAE: *Hydrodictyon* (*hydro*, water; *dictyon*, net). Fig. 31.

The number of species are few, the commonest, *H. reticulatum*, having a world-wide distribution though it occurs but rarely in each locality. It is a hollow, free-floating, cylindrical network closed at either end and up to 20 cm. in length. The individual coenocytic cells are multinucleate and are arranged in hexagons or pentagons to form the net. The chloroplast is reticulate with numerous pyrenoids, though in the young uninucleate cells there is but a simple parietal chloroplast which later becomes spiral and then reticulate. *H. africanum* and *H. patenaeforme* develop into saucer-shaped nets, the former with spherical cells up to 1 cm. diameter which may become detached and lie on the substratum looking like pearls. The other species is composed of cells which may grow up to 4 cm. long by 2 mm. in diameter. Asexual reproduction in *H. reticulatum* is by means of numerous uninucleate zoospores which swarm in the parent cell about daybreak and then come together to form a new coenobium which is subsequently liberated, further growth being brought about by elongation of the coenocytic cells. It is interesting to note that the arrangement of the daughter cells in the parent coenocyte agrees with the mechanical laws for obtaining the greatest rigidity with the maximum economy of space.

Asexual reproduction is unknown in *H. africanum* and *H. patenae-*

forme. Sexual reproduction in all three species is isogamous in character and the plants are monoecious. In *H. patenaeforme* the zygote is motile for a short time, but in the other two species it is always non-motile. At germination the zygote enlarges and divides by meiosis into four biflagellate swarmers which first come

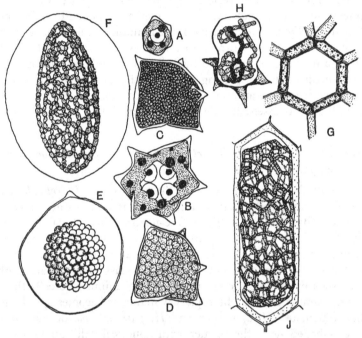

Fig. 31. *Hydrodictyon*. A–F, development of young net of *H. patenaeforme* from the zygote. A, young polyhedron. B, older polyhedron with four nuclei. C, protoplasm granular just before zoospore formation. D, "pavement" stage. E, zoospores rounding off and wall of polyhedron expanding to form vesicle. F, fully formed net still enclosed in vesicle. (A–E × 250, F × 175.) G, portion of mature net of *H. reticulatum*. H, polyhedron and young net of *H. reticulatum*. J, *H. reticulatum*, formation of net in parent cell from zoospores. (A–F, after Pocock; G, H, after Oltmanns; J, after Fritsch.)

to rest and then develop into polyhedral cells. After resting for a period these divide to produce zoospores; the food material in the angular thickenings of the polyhedrons is used up and all the swarmers are finally liberated in a vesicle, in which, after a period of motility, they come together to form a new coenobium. The vegetative plant is therefore haploid and its development is probably one of the most remarkable that is to be found among the fresh-water algae. *Pediastrum* is a very poor indication of what the ultimate development of this type of thallus construction could be, and this provides a problem at present unsolved, namely, the absence of any intermediate morphological stage between *Pediastrum* and *Hydrodictyon*. Further increase in the size of colony is probably impossible for purely mechanical reasons. The fact that the cells are coenocytic also indicates that the siphonaceous habit must have arisen several times in the course of evolution. *Hydrodictyon* is essentially a collection of a number of individual coenocytic plants because it has arisen as a result of the fusion of a number of swarmers. *Volvox*, on the other hand, must be regarded as a single plant composed of a number of cells connected by strands because it arises from a single zygote or asexual cell. Gamete and zoospore production respectively can be obtained in *Hydrodictyon* by varying the external conditions artificially. For example, if plants are grown in weak maltose solutions in bright light or in the dark and are then transferred to distilled water, zoospores will develop under the first set of conditions and gametes under the second.

COELASTRACEAE: *Scenedesmus* (*scene*, rope; *desmus*, fetter). Fig. 32.

The planktonic colonies are composed of four, eight or, more rarely, sixteen cells. The two end cells of the chain may differ in

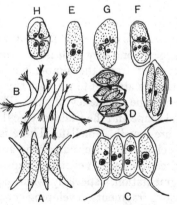

Fig. 32. *Scenedesmus*. A, *S. acuminatus*. B, *S. acuminatus* with mucilage bristles. C, *S. quadricauda*. D, *S. quadricauda* reproducing. E–I, stages in the formation of daughter coenobia in *S. quadricauda*. (After Fritsch.)

shape from the others and often have processes which are elabora-
tions of the mucilaginous cell envelope: these processes are probably to be correlated with the planktonic mode of life, whilst tufts of
bristles performing the same function and similar to those of
Pediastrum are also recorded.

It should be evident from the preceding descriptions that the
Chlorococcales represent a number of very diverse types, some of
which may have indications of distant relationships whilst there
are others whose relationships are extremely vague: a recent paper
even describes some oogamous members.

REFERENCES

Chlorococcum. BOLD, H. C. (1930–1). *Bull. Torrey Bot. Club,* **57,** 577.
Chlorochytrium. BRISTOL, B. M. (1919). *J. Linn. Soc. (Bot.)* **45,** 1.
Chlorella. GRINTZESCO, J. (1903). *Rev. Gén. Bot.* **15,** 5.
Pediastrum. HARPER, R. A. (1918). *Proc. Amer. Phil. Soc.* **57,** 375.
Hydrodictyon. MAINX, F. (1931). *Arch. Protistenk.* **75,** 502.
Chlorochytrium. PALM, B. T. (1932). *Rev. Alg.* **6,** 337.
Hydrodictyon. POCOCK, M. A. (1937). *Trans. Roy. Soc. S. Afr.* **24,** 263.
Chlorococcum. PUYMALY, A. DE (1924). *Rev. Alg.* **1,** 107.
Scenedesmus. SMITH, G. M. (1914). *Arch. Protistenk.* **32,** 278.

ULOTRICHALES

ULOTRICHACEAE: Ulothrix (ulo, shaggy; *thrix,* hair). Fig. 33.

The unbranched filaments are attached to the substrate by means
of a modified basal cell which frequently lacks chlorophyll, but even
though attached at first the plants sometimes become free-floating.
Under unfavourable conditions, e.g. nutrient deficiency, rhizoids
may grow out from the cells or else the filaments become branched.
This behaviour suggests one way at least in which the branched
habit may have evolved from the simple filament, in this case
probably representing an attempt to increase the absorbing surface
in order to counteract the deficiency of salts. The cells vary considerably in size and shape and the walls may be thick or thin; if the
former, then they are usually lamellate. There is a single chloroplast which forms a characteristic circular band around the whole or
most of the cell circumference. Vegetative reproduction can take
place through fragmentation, especially when conditions are unfavourable, the various fragments developing conspicuous rhizoids.

Swarmers are formed from all the cells of the filament except the
attachment cell, but they usually appear first at the apex of the

filament and then successively in the other cells. They are liberated through a hole in the side of the cell into a delicate vesicle, and the subsequent bursting of this vesicle frees the swarmers, all of which

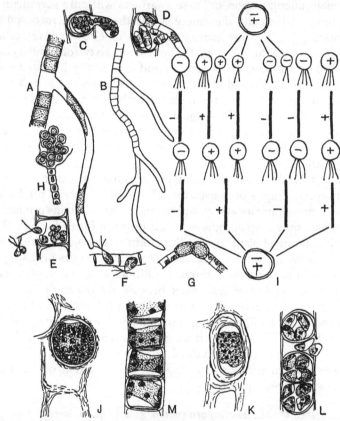

Fig. 33. *Ulothrix zonata*. A, B, rhizoid formation. C, liberation of swarmers into vesicle (× 375). D, germination of aplanospores in the cell (× 250). E, liberation of gametes (× 375). F, escape of zoospores (× 375). G, akinetes of *U. idiospora*. H, palmelloid condition. I, schema to illustrate the different types of filaments and swarmers. J, K, aplanospores (× 400). L, zoospore formation (× 400). M, banded chloroplasts in a portion of the vegetative filament (× 400). (A, B, I, after Gross; C–F, J–M, after West; G, H, after Fritsch.)

are positively phototactic. Three types of swarmer are to be found. (*a*) Quadriflagellate macrozoospores, of which two, four or eight are produced per cell. After a motile period these become broader than they are long, attach themselves to a suitable substrate, and then a rhizoid and filament grow out opposite each other in a plane at right angles to the longitudinal axis of the original zoospore. (*b*) Each cell produces four, eight, sixteen or thirty-two bi- or quadri-flagellate microzoospores. These swarmers will only germinate at low temperatures and then more slowly than the macrozoospores, producing a somewhat narrow filament or else forming resting spores. (*c*) Eight, sixteen, thirty-two or sixty-four biflagellate gametes are produced in each cell and are usually liberated soon after daybreak. The adult plants are usually dioecious, and after fusion of the gametes has taken place (parthenogenesis is said not to occur) the quadriflagellate zygote forms a resting zygospore. This germinates after 5–9 months giving rise to four or sixteen aplano-spores, and as meiosis occurs during their production the adult plants are haploid.

It is said that there are six types of adult filament: + and − strains producing + or − gametes only, + and − strains producing both + or − gametes and zoospores and + or − strains producing zoospores only. Aplanospores, when they are formed, may either develop into new plants or else they form a temporary "palmella" state. Akinetes are also recorded. The genus appears primarily in winter or spring, and the optimum conditions would seem to include either cold weather or cold water because the plants die down in summer. The genus is well represented in both fresh and salt waters, *U. zonata* and *U. flacca* being common species respectively of the two habitats. The nearly related genus *Schizomeris*, in which the filaments have some longitudinal divisions, may be considered as representing an intermediate stage in the evolution of the more foliaceous forms, e.g. *Ulva*.

*MICROSPORACEAE: *Microspora* (*micro*, small; *spora*, seed). Fig. 34.

This genus is sufficiently distinct from the preceding one to warrant its inclusion in a separate family. The plants are free-floating when mature and consist of unbranched threads, the cells of which have walls of varying thickness, the thicker walls showing some stratification. In many species the cell wall is in two over-lapping halves held in place by a delicate inner or outer membrane, and it is because of this type of structure that the threads readily fragment into H pieces. In ordinary cell division growth is brought about by the introduction of new H pieces. The parietal chloroplast is often reticulate or else forms an irregularly thickened band, and, although there are no pyrenoids, the genus is characterized by an abundance of starch in the threads. One to sixteen biflagellate (quadriflagellate in one species) zoospores are formed in each mother cell and are liberated by the thread fragmenting into H

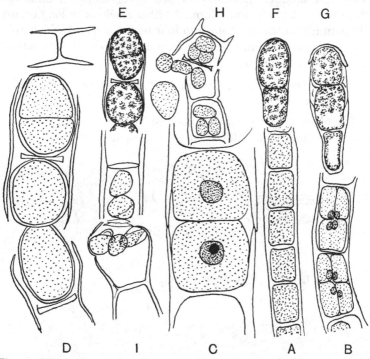

Fig. 34. *Microspora amoena*. A, portion of thread. B, early cleavage in swarmer formation. C, two young cells (× 745). D, akinete formation (× 550). E, formation of aplanospores. F, G, stages in germination of aplanospores. H, liberation of zoospores. I, zoospores (× 745). (C, D, I, after Meyer; rest after Fritsch.)

pieces or else by gelatinization of the cell walls. There may perhaps be biflagellate gametes, but fusion between swarmers has only been seen in one species, whilst in another species gametes possessing somewhat unequal flagellae have been recorded. This fact is extremely interesting and, if true, would make a reorientation of ideas about this genus essential. *Microspora* exhibits considerable variation, particularly in a xanthophycean direction, and in many characters it overlaps *Tribonema* (cf. p. 117). For this reason it is not impossible that the filamentous Xanthophyceae may be derived via a form such as *Microspora* from an ulotrichaceous filament. On this view *Ulothrix* and *Tribonema* cannot be regarded as end-phases in separate lines of evolutionary development. Any cell of *Microspora* can produce aplanospores instead of motile bodies, and akinetes are also frequently formed, either singly or in long chains. On germination these divide into four bodies, each of which gives rise to a new filament. This genus, like *Ulothrix,* is also confined to the winter or spring months.

CYLINDROCAPSACEAE: *Cylindrocapsa* (*cylindro*, cylinder; *capsa*, box). Fig. 35.

The filaments, which are unbranched, are attached at the base by means of a gelatinous holdfast, and when young each thread is com-

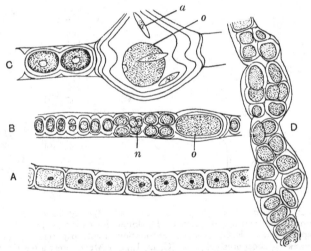

Fig. 35. *Cylindrocapsa*. A, vegetative filament. B, thread with young antheridia (*n*) and young oogonium (*o*). C, fusion of gametes. *a* = antherozoid, *o* = ovum. D, old mature filament. (After Fritsch.)

posed of a single row of elliptical cells with thick stratified walls, the whole being enclosed in a tubular sheath. In older filaments, however, the cells divide longitudinally, usually into pairs, and the gelatinous nature of the threads suggests how the genus *Monostroma* may have evolved, although *Cylindrocapsa* itself has developed much farther, especially in its mode of sexual reproduction, because it is anisogamous with elongate antherozoids and large round ova.

MONOSTROMACEAE: *Monostroma* (*mono*, single; *stroma*, layer). Figs. 36, 37.

The thallus develops as a small sac, which in most species ruptures very early to give a torn plate of cells one layer in thick-

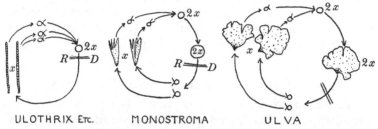

ULOTHRIX ETC.   MONOSTROMA   ULVA

Fig. 36. Diagram to illustrate the three different types of life cycle found in the Ulotrichales. *RD* = place of reduction division in life cycle.

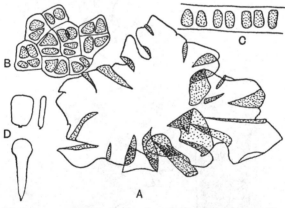

Fig 37. *Monostroma crepidinum*. A, plant (× ⅔). B, cells of thallus (× 200). C, transverse section of thallus (× 200). D, *M. Lindaueri*, plants (× ⅔). (After Chapman.)

ness, the cells often being arranged in groups of two or four. In *M. Grevillei* the thallus only ruptures in the adult stage, and traces of the original tubular form can also be seen in adult plants of *M. Blytii*, whilst in *M. Lindaueri* the sac appears to remain entire.

The plants are dioecious in respect of sexual reproduction and several of the species exhibit anisogamy. Biflagellate gametes from separate plants fuse to give a non-motile zygote which then increases in size and after some months undergoes meiosis and forms zoospores. The macroscopic plants are thus all haploid and the diploid generation is only represented by the enlarged zygote. In this respect it is sharply differentiated from the genera *Ulva* and *Enteromorpha*, and it possibly has only a distant relationship with them. Each zoospore from the zygote divides to give eight peripherally arranged cells with a central cavity and this then develops slowly into a sac. The genus is more widespread than is perhaps suspected from the literature, frequenting both saline and fresh waters.

*ULVACEAE: *Ulva* (Latin for a marsh plant). Figs. 36, 38.

The thallus, which is composed of two layers and is therefore distromatic, develops from a single uniseriate filament that sub-

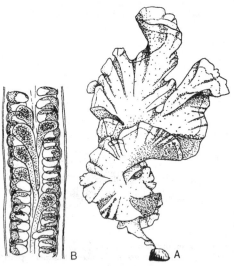

Fig. 38. *Ulva lactuca.* A, plant. B, transverse section of thallus. (After Oltmanns.)

sequently expands by lateral divisions, but there is usually no hollow sac, though exceptions to this are found in *U. Linza* and *U. rhacodes.* The plant is attached at first by a single cell, but later multinucleate rhizoids grow down from the lower cells and a basal attachment disk is formed which may persist throughout the winter, new plants arising from it in the spring. Detached fragments are another frequent means of forming new thalli, whilst normal asexual reproduction is by means of quadriflagellate zoospores. In sexual reproduction, which occurs in plants other than those producing zoospores, fusion takes place between isogametes from separate plants which have been described as + and −. The gametes may fuse in pairs or they may fuse into "clumps", and whilst they are positively phototactic before fusion, the zygote is negatively phototactic, and this change in behaviour causes it to descend on to a suitable substrate. Hartmann (1929) has shown that in certain cases there may be relative sexuality among gametes from different plants, the sex of the older and weaker gametes becoming changed. Meiosis takes place at zoospore formation and there is a regular alternation of diploid and haploid generations, both indistinguishable morphologically, and when this life history is compared with that of *Monostroma* the essential differences are immediately apparent (cf. fig. 36). The plants occur in saline or fresh water and become particularly abundant when the waters are polluted by organic matter or sewage.

*ULVACEAE: *Enteromorpha* (*entero*, entrail; *morpha*, form). Figs. 36, 39.

The plants of this genus also commence life as uniseriate filaments which soon become multiseriate and tubular. Like *Ulva*, many of the species are attached by means of rhizoids, but there are also a number of forms, especially on salt marshes (cf. p. 330), which are free-floating for the whole or part of their life cycle. Growth of the thallus is either intercalary or else through the divisions of an apical cell. Asexual reproduction is by means of zoospores, and as meiosis takes place at their formation the life cycle is identical with that of *Ulva* because morphologically similar haploid plants are known. The first division of the germinating zoospore is transverse, the lower segment forming an embryonic rhizoid. The sexual haploid plants are dioecious, usually with isogamous reproduction, the gametes commonly being liberated around daybreak. Anisogamy has been found by Kylin (1930) in *E. intestinalis* where the male gamete is small with but a rudimentary pyrenoid. The motile phase of the gametes is short, lasting about 24 hours, whilst the zygote may also remain motile for 1 hour, although the first division of the zygote usually takes place after several days' dormancy.

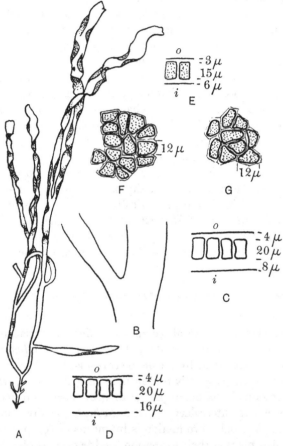

Fig. 39. *Enteromorpha intestinalis* f. *flagelliformis*. A, portion of plant. B, origin of branch of same showing basal constriction. C, D, E, transverse sections from near base, middle and apex of thallus. *o* = outside, *i* = inside of tube. F, G, cells of thallus. (Original.)

Parthenogenetic development of gametes has been recorded for *E. clathrata*, and this presumably results in new sexual plants. As yet no evidence of relative sexuality has been found among the gametes of this genus.

*PRASIOLACEAE: *Prasiola* (*prasio*, green). Fig. 40.

The young unbranched filament, which is known as the "*Hormidium*" stage, consists of a single row of flat cylindrical cells with thick walls which frequently possess striations. Later on the cells divide longitudinally and produce a thin expanded thallus, known as the "*Schizogonium*" stage, which tapers to the base. The cells of

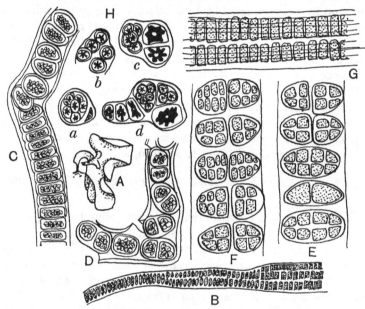

Fig. 40. *Prasiola*. A, plant of *P. crispa*. B, "*Schizogonium*" stage of *P. crispa* forma *muralis*. C, D, "*Hormidium*" stage of *P. crispa* f. *muralis* with akinetes. E, development of macrogametes in *P. japonica* (×665). F, development of microgametes in *P. japonica* (×665). G, *P. crispa*, membrane striations in "*Schizogonium*" stage (×650). H (a–d), formation of aplanospores in akinetes and young plants. (A, B, after Fritsch; C, D, H, after Oltmanns; E–G, after Knebel.)

the mature thallus are often arranged in fours and possess axile stellate chloroplasts, whilst another feature is the presence of short rhizoids that may occur in the stalk-like portion or else growing out from the marginal cells. In the juvenile filament reproduction takes place by means of fragmentation as a result of the death of isolated cells, whilst in the older, more leafy thallus, "buds" can arise from the margins. Sometimes the cells produce large, thick-walled akinetes that germinate to form aplanospores from which new plants arise. In *P. japonica* sexual reproduction is brought about by macro- (sixteen per cell) and microgametes (sixty-four per cell) that are both produced on the same plant so that this species, at least, is anisogamous. The shape of *P. crispa* has been shown to vary considerably with the habitat, the optimum conditions being those where there is an abundant supply of nitrogen, such as may be found in areas occupied by bird colonies. The genus, which is generally absent from the tropics and subtropics, is represented by saline, fresh-water or subaerial species, the latter being tolerant towards considerable desiccation and temperature changes. This resistance is attributed to the lack of vacuoles in the cells and also to the high viscosity of the protoplasm. Water supply appears to be the principal factor limiting successful development, especially in the subaerial species. Some authors consider that the genus is characterized sufficiently to warrant removal from the Ulotrichales, but such a change does not really seem to be justified.

SPHAEROPLEACEAE: *Sphaeroplea* (*sphaero*, sphere; *plea*, full). Fig. 41.

This genus is widely distributed, being most abundant on ground that is periodically flooded by fresh water. The long, free, unbranched filaments consist of elongated coenocytic cells containing one to seventy annular parietal chloroplasts. These latter have denticulate margins and occupy the periphery of disks of cytoplasm, the disks being separated from each other by vacuoles, although occasionally they may come together to form a diffuse network. Each disk normally possesses one or two nuclei in its cytoplasm. In most of the species the septa develop as ingrowths, though in *S. Africana* these are replaced by a series of processes which appear to be comparable to the strands of a *Caulerpa* (cf. p. 91), but as they sometimes fail to meet at the centre the coenocytes may be continuous.

Vegetative reproduction is secured by means of fragmentation and there is apparently no asexual reproduction. In sexual reproduction although the cells do not change in shape, nevertheless both oogonia and antheridia are formed singly or in series, the plants being either monoecious or dioecious. In the formation of oogonia the annular chloroplasts first become reticulate and then the ova are formed without any nuclear divisions being involved. In *Sphaeroplea annulina* the ova are non-motile but in *S. cambrica* they are biflagellate, and so it may be argued that the motionless egg has been evolved from the motile one by loss of flagellae. In the antheridia, on the other hand, the nuclei undergo division and

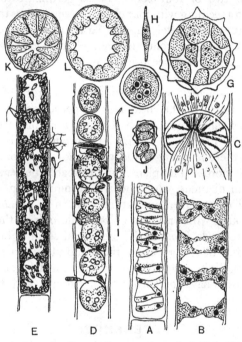

Fig. 41. *Sphaeroplea.* A, *S. annulina*, portion of thallus. B, *S. annulina*, chloroplast. C, structure of septum in *S. Africana* (× 375). D, female plant with ova and antherozoids. E, male plant. F, young zygote. G, zygote with thickened wall. H, I, young gametophytes. J, spores emerging from zygote. K, L, *S. Africana*, transverse sections across the septa (× 375). (A–C, K, L, after Fritsch; D–J, after Oltmanns.)

numerous elongated narrow antherozoids are formed which are liberated through small holes, subsequently penetrating the oogonia through similar perforations. The fertilized ovum (*oospore*) becomes surrounded by a hyaline membrane, and then inside this two new membranes are laid down, after which the first one disappears. The new external membrane is ornamented and the contents of the oospore are now a brick red. Germination stages are only known

56 CHLOROPHYCEAE

for a few species, and in such cases the oospores may remain dormant for several years before they produce one to four biflagellate swarmers which very soon come to rest and then grow into new plants. On germination the zoospores do not always separate and so one gets a four- or eight-flagellate *synzoospore* depending on whether it is composed of two or four zoids. These germinate to a fourfold seedling or to a seedling with four claws. In some cases the swarmers in the oospore are completely suppressed and a new filament develops directly, this type of reproduction being known as *azoosporic*. The adult plants are haploid because meiosis is known to take place at the segmentation of the oospore.

Primitive characters, which seem to be a feature of the genus, are the numerous ova, the entire lack of specialized organs for producing the sexual reproductive bodies and a simple form of zygote germination, whilst in *S. tenuis* the reproduction is even more primitive as there is strong evidence to show that both kinds of gametes are motile. The plant must probably be regarded as an Ulotrichacean filament, which, whilst becoming non-septate, has still retained many primitive features, and in *S. annulina* cells are frequently found with only one or two chloroplasts thus showing a gradation towards *Ulothrix*. There would seem to be very little justification for following some authors and placing it in either the Siphonales or Siphonocladiales, though it must be admitted that *S. Africana* does have some features in common with those of the Siphonocladiales.

REFERENCES

Enteromorpha. BLIDING, C. (1933). *Svensk bot. Tidskr.* **27**, 233.
Ulvaceae. CARTER, N. (1926). *Ann. Bot., Lond.,* **40**, 665.
Sphaeroplea. FRITSCH, F. E. (1929). *Ann. Bot., Lond.,* **43**, 1.
Ulothrix. GROSS, I. (1931). *Arch. Protistenk.* **73**, 206.
Enteromorpha. HARTMANN, M. (1929). *Ber. dtsch. bot. Ges.* **47**, 485.
Prasiola. KNEBEL, G. (1935-6). *Hedwigia,* **75**, 1.
Monostroma. KUNIEDA, H. (1934). *Proc. Imp. Acad. Tokyo,* **10**, 103.
Enteromorpha KYLIN, H. (1930). *Ber. dtsch. bot. Ges.* **48**, 458.
Ulothrix. LIND, E. M. (1932). *Ann. Bot., Lond.,* **46**, 711.
Microspora. MEYER, K. (1913). *Ber. dtsch. bot. Ges.* **31**, 441.
Sphaeroplea. PASCHER, A. (1939). *Beih. bot. Zbl.* **59**, Abt. A., 188.
Microspora. STEINECKE, F. VON (1932). *Bot. Arch.* **34**, 216.
Prasiola. YABE, Y. (1932). *Sci. Rep. Tokyo Bunrika Daig.* **1**, 39.

OEDOGONIALES

Oedogonium (*oedo*, swelling; *gonium*, vessel). Figs. 42, 43.

The three genera, *Oedogonium*, *Oedocladium* and *Bulbochaete*, which comprise this order were at one time classed as a separate group, the Stephanokontae. Under the new scheme of classification, however, they must be regarded, together with the other members of the old Isokontae, as forming the Chlorophyceae.

In *Oedogonium* the thallus consists of long unbranched threads which are attached when young, though later they become free-floating, whilst in the other two genera the filaments are commonly branched. Each cell possesses a single nucleus together with an elaborate reticulate chloroplast containing numerous pyrenoids. The cell wall contains, according to some workers, an outer layer of chitin, and if they are correct this is of great interest because chitin is essentially an animal substance. The chromosomes of *Oedogonium* are especially interesting among those of the algae in that they have thickened dark segments at intervals along their length. Vegetative cell division is so peculiar and characteristic that many accounts of the process have appeared. A thickened transverse ring, which develops near the upper end of the cell, first enlarges and then invaginates, the much thickened wall being pushed into the interior of the cell. Nuclear division now takes place near this end of the cell and a septum is laid down between the two daughter nuclei. Next, the outer parent cell wall breaks across at the ring and the newly formed membrane stretches rapidly now that the pressure is released—a matter of about 15 min.—so that a new cell is interposed between the two old portions. The new transverse septum becomes displaced by differential growth of the two daughter cells so that it finally comes to rest just below the fractured parent wall, and it is also evident that the new longitudinal wall of the upper cell is almost entirely composed of the stretched membranous ring. The old walls form a cap at one end and a bottom sheath at the other, and as successive divisions always occur at the upper end of the same cells, a number of caps develop there and give the characteristic striated appearance to some of the cells. This method of growth in *Oedogonium* may be either terminal or intercalary, but in the other two genera, as each cell can only divide once, there is usually only a single cap. This peculiar mode of

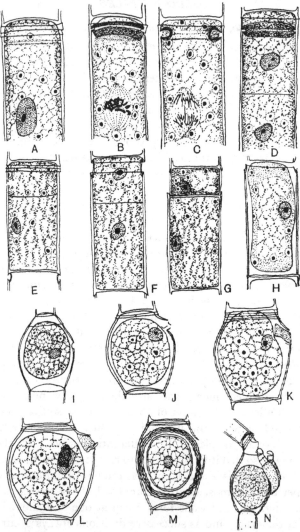

Fig. 42. *Oedogonium.* A–G, stages in cell division in *Oe. grande* (×526). B, C, formation of ring. F, G, expansion of ring to form new cell. H, formation of aplanospore in *Oe. Nebraskense.* I, *Oe. ciliatum,* position of antherozoid 2 hours after entering egg. J–M, stages in fertilization of ovum of *Oe. Americanum.* K, entrance of sperm. L, fusion of gamete nuclei. M, zygote. N, *Oe. Kurzii,* dwarf male (×175). (A–M, after Ohashi; N, after Pringsheim.)

division is unique, and although there is no trace of its ancestry its constancy suggests that the group terminates a line of evolutionary development. Vegetative reproduction commonly occurs by means of fragmentation, whilst asexual reproduction is

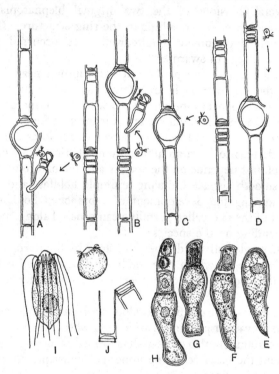

Fig. 43. *Oedogonium.* A, idioandrosporous nannandrous filament. B, gynandrosporous nannandrous filament. C, dioecious macrandrous filament. D, monoecious filament. E–H, stages in development of dwarf male plant (×400). I, antherozoid (×480). J, escape of zoospore (×138). (A–D, after Mainx; E–I, after Ohashi; J, after West.)

secured through akinetes or multiflagellate zoospores, the formation of the latter being said to depend on the presence of free carbon dioxide in the water. The flagellae, which may have one or two rings of granular blepharoplasts at their base, form a circular ring around an anteriorly situated beak-like structure. This is the typical oedogonian swarmer, one of which is produced by each cell, and there are two theories that have been put forward to explain its origin:

(*a*) The group arose independently from flagellate organisms which possessed a ring of flagellae. If this is true then there could be no real connexion with the other members of the Chlorophyceae.

(*b*) Several divisions of the two original blepharoplasts and flagellae took place, thus resulting in the ring structure. If this is correct then development might well have occurred from a Ulotrichalean type of swarmer.

When the zoospore is ripe the cell wall ruptures near the upper end and the swarmer is liberated into a delicate mucilaginous vesicle, but this soon disappears, thus allowing the zoospore to escape. After remaining motile for about an hour the anterior end becomes attached to some substrate and develops into a holdfast, or else the zoospore flattens to form an almost hemispherical basal cell. The type of holdfast depends on the species and the nature of the substrate, a smooth surface inducing a simple holdfast and a rough surface inducing the development of a branched holdfast. Development of the one-celled germling can proceed along one of two lines, depending on the species:

(*a*) The single cell divides near the apex by the normal method described above, in which case the basal daughter cell persists as the attachment organ and the upper cell goes on to form the new filament.

(*b*) The apex of the cell first develops a cap and then a cylinder of protoplast grows out pushing it aside, and when the protoplast has reached a certain length a cross-wall is formed at the junction of the cylinder and the basal cell. The upper cell subsequently develops along the normal lines.

Sexual reproduction is by means of an advanced type of oogamy, the development of sex organs being assisted by an alkaline *p*H and some nitrogen deficiency. In some of the species the oogonia and antheridia are produced on the same plant (*monoecious* forms): in other species the oogonia and antheridia appear on different filaments which are morphologically alike (*dioecious homothallic* forms). The species belonging to both these groups are termed *macrandrous* because the male filament is normal in size. There is a third group of species in which the male filament is much reduced and forms dwarf male plants. Such species are *dioecious* and *hetero-*

thallic and they form the *nannandrous* group. The dwarf males arise from motile *androspores* which are formed singly in flat discoid cells, the *androsporangia*, produced by repeated divisions of ordinary vegetative cells. The androspores may be formed either in the oogonial filament—*gynandrosporous* species—or on other filaments that do not bear oogonia—*idioandrosporous* species (fig. 43). In shape and structure the androspores are small editions of the zoospores, and after swimming about they settle on the wall of the oogonium or on an adjacent cell and germinate into a small male plant which is composed of a rhizoidal holdfast with one or two flat antheridia above, though in some cases only one antheridial cell without any rhizoidal portion is formed. Usually two antherozoids are freed from each antheridium into a delicate vesicle which later dissolves. The antherozoids are also like small zoospores, and if they fail to enter an ovum immediately they may remain motile for as long as 13 hours. In the macrandrous monoecious species the antheridia are usually to be found immediately below the oogonia where they arise by an ordinary vegetative division in which the upper cell subsequently continues to divide rapidly, thus producing a series of from two to forty antheridia. The antheridia frequently develop one day later than the oogonia, thus ensuring cross-fertilization.

The oogonia are enlarged spherical or ellipsoidal cells arising by one division in which the upper segment forms the oogonium and the lower a support cell, or else the latter subsequently divides to give antheridia. In some species the lower cell may also become an oogonium so that one can find a series of oogonia on one filament. Each oogonium contains one ovum with a colourless receptive spot situated opposite to the opening in the oogonium wall from which a small quantity of mucilage is extruded. The opening is either a very small pore, formed by gelatinization of a tiny papilla, or else a slit, but in either case there is an internal membrane forming a sort of conduit to the ovum. After fertilization the oospore often becomes reddish in colour and develops a thick membrane which is usually composed of three layers. At germination the protoplast divides into four segments, which may each develop flagellae and escape as zoospores, or else they function as aplanospores that later give rise to zoospores. Meiosis takes place at the germination of the zygote so that the adult filaments are haploid. In one species it has been definitely established that two of the zygote segments ultimately develop into male plants and two into female plants. Zygote germination without meiosis is not uncommon, in which case it gives rise to what are presumably large diploid swarmers, and these develop into abnormally large threads that are always female. Oogonia appear on these diploid filaments and can be fertilized, but the fate of the zygote is unknown.

It remains to discuss the possible origin of the androspores, and there are two hypotheses that may be considered:

(a) The androspore is equivalent to the second and smaller type of asexual zoospore, such as those found in *Ulothrix*, but in the Oedogoniales they can no longer give rise to normal filaments. On this view the nannandrous forms are the more primitive, the macrandrous having been derived by the androsporangium acquiring the capacity to produce antheridia immediately and hence never appearing. (b) The androspore is equivalent to a prematurely liberated antheridial mother cell which subsequently undergoes further development. On this view the macrandrous species are the more primitive. West (1912) considered that the dwarf males were to be regarded as reduced from normal male filaments, for in one species the male plants are intermediate in size. At present there does not appear to be any very convincing evidence in support of either theory.

REFERENCES

Gussewa, K. (1931). *Planta*, **12**, 293.
Mainx, F. (1931). *Z. Bot.* **24**, 481.
Ohashi, H. (1930). *Bot. Gaz.* **90**, 177.
Spessard, E. A. (1930). *Bot. Gaz.* **89**, 385.
West, G. S. (1912). *J. Bot.* **50**, 321.

CHAPTER IV

CHLOROPHYCEAE (CONT.)

CHAETOPHORALES, SIPHONOCLADIALES, SIPHONALES

CHAETOPHORALES

A family in which the fundamental structure is the possession of both a basal and erect system, this type of thallus being known as heterotrichous (cf. p. 263). In some of the genera, however, reduction has taken place and only the basal or erect system is now represented.

*PLEUROCOCCACEAE: *Pleurococcus* (*pleuro*, box; *coccus*, berry). Fig. 44.

The systematic position of this alga has varied considerably. By some authors it has been placed in the Chlorococcales whilst others

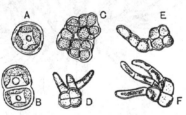

Fig. 44. *Pleurococcus Naegelii.* A, single cell. B, single-celled colony. C, normal colony. D–F, thread formation. (After Fritsch.)

have placed it in a special group, the Pleurococcales, but as the alga can occasionally develop branched threads there would seem to be evidence for regarding it as a much reduced member of the Chaetophorales. There are, it is true, almost equally sound arguments for the other systematic treatments of the genus, and its place at present must be largely a matter of opinion. *Pleurococcus* is terrestrial and forms a green coat on trees, rocks and soil, growing in situations where it may have to tolerate prolonged desiccation. The cells, which are globose in shape and occasionally branched, are single, or else as many as four may be united into a group. Under certain cultural conditions branching may be copious. Each cell contains one chloroplast and there are no pyrenoids. The sole method of reproduction is through vegetative division in three planes when one may find up to fifty cells in a group. There is probably only one species, *P. Naegelii*, all the other so-called species being reduced or modified forms of other algae. The resistance of the cells to desiccation is aided by a highly concentrated cell sap and a capacity to imbibe water directly from the air.

*CHAETOPHORACEAE: *Draparnaldia* (after J. P. R. Draparnaud). Fig. 45.

The plants, which are confined wholly to fresh water, are represented principally by the aerial system, the prostrate system being

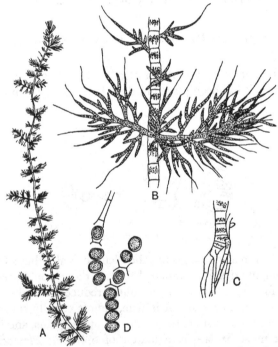

Fig. 45. *Draparnaldia.* A, portion of plant (× ⅔). B, same enlarged. C, rhizoids in *D. plumosa.* D, aplanospores of *D. glomerata.* (A, B, D, after Oltmanns; C, after Fritsch.)

entirely absent or else greatly reduced. The young plant is originally attached by means of a much reduced prostrate system together with rhizoids from one or two basal cells. The thallus, which is often invested by a gelatinous matrix of pectins, possesses a main axis composed of large barrel-shaped cells, each containing a small entire or reticulate chloroplast and several pyrenoids. This axis is primarily for support, and it bears much branched laterals that normally grow out in tufts, the short cells composing the laterals being almost wholly filled by one entire chloroplast containing a single pyrenoid. The apices of these branches, which perform the functions of assimilation and reproduction, are often prolonged into a hair. In some species rhizoids develop at the base of the branches and grow downwards thus clothing the main axis with a pseudo-cortex, but normal growth is generally restricted to a few cells of the thallus. When grown in culture with increased carbon dioxide or additional nitrate the plants take on a form very like that of *Stigeoclonium* (cf. below). Asexual reproduction is by means of quadriflagellate macrozoospores, one to four being produced in each cell. These, after swarming for a few minutes, settle, and germinate into a short filament which already possesses a hair at the four- or five-celled stage when it commences to put out rhizoids. Sexual reproduction is secured by means of quadriflagellate microswarmers or isogametes which fuse whilst in an amoeboid state, though these gametes may also develop parthenogenetically. The behaviour of the microswarmers demands further investigation as it does not seem to be clearly understood, nor has it been determined whether the plants are haploid or diploid. In *Draparnaldia glomerata* the nature of the swarmer is controlled by the pH of the medium, microswarmers being formed under alkaline conditions and macrozoospores under neutral or acid conditions.

*CHAETOPHORACEAE: *Stigeoclonium* (*stigeo*, sharp pointed; *clonium*, branch) (*Myxonema*). Fig. 46.

Many species are heterotrichous and the plants are frequently enclosed in a broad watery gelatinous sheath. The chloroplast is band-like and often does not fill the entire cell, especially in the older parts of the thallus. The aerial part bears branches that terminate in a colourless hair, the degree and nature of the branching depending upon illumination, nutrition and the rate of water flow. There is no localized area for cell division in the aerial portion, but in the creeping system only the apical cells are meristematic. The prostrate system may be (*a*) loosely branched, (*b*) richly and compactly branched or (*c*) a compact disk, but the more developed the basal portion the less elaborate is the aerial and vice versa. Vegetative reproduction is by means of fragmentation, whilst sexual and asexual reproduction are the same as in *Draparnaldia*, except that there is only one macrozoospore

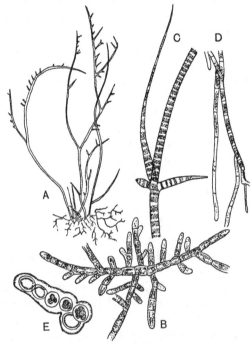

Fig. 46. *Stigeoclonium.* A, plant of *S. tenue.* B, basal portion of *S. lubricum.* C, aerial position of *S. protensum.* D, rhizoids in *S. aestivale.* E, palmelloid state. (A–C, E, after Oltmanns; D, after Fritsch.)

produced per cell. In two species, however, a third type of biflagellate swarmer is known, and hence reproduction in these species is comparable to that found in *Ulothrix* (cf. p. 46). These extra swarmers, which are probably the true gametes, are few in number but fusion between them is rare, probably because the plants are dioecious. In general the microswarmers seem to have taken over the function of the sexual biflagellate gametes. The zygote is said to germinate to zoospores, and these then give rise to the germlings in which the erect filament arises first and the prostrate portion subsequently or vice versa. By increasing the osmotic pressure or by adding toxic salts to the environment the thallus passes into a palmelloid state, whilst under other conditions akinetes can be formed. The plants are confined to well-aerated fresh water though they have also been found growing on fish living in stagnant water, but in these cases the movements of the fish presumably provide adequate aeration.

TRENTEPOHLIACEAE: *Gongrosira* (*gongro*, excrescence; *sira*, chain). Fig. 47.

A genus which lives on stones and the shells of gastropods that are to be found in fresh and salt water, although there is one species that is terrestrial. The cushions or plates are frequently

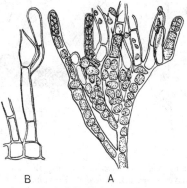

B   A

Fig. 47. *Gongrosira.* A, portion of *G. circinnata* showing formation of zoospores. B, dehisced sporangium of *G. stagnalis.* (After Fritsch.)

lime encrusted and form a tough green stratum with a base that is composed of one or more layers of cells which give rise to dense, erect, branched filaments. The sporangia are usually borne terminally on these erect threads, and in some species they can even be distinguished morphologically by their greater size, although generally they do not differ from the vegetative cells. The sporangia produce biflagellate zoospores, and any of these which are not able to escape become converted into aplanospores. Biflagellate isogametes develop from the lower cells of the thallus, whilst the prostrate portion can also give rise to akinetes.

TRENTEPOHLIACEAE: *Cephaleuros* (*cephal*, head; *euros*, broad).
    Fig. 48.

These grow as epiphytes and parasites on and in the leaves of various phanerogams. The plants are composed of one or more branched interwoven threads from which vertical filaments arise that bear clusters of stalked sporangia very like those of *Trente-pohlia*. Some species bear sterile erect filaments that terminate in

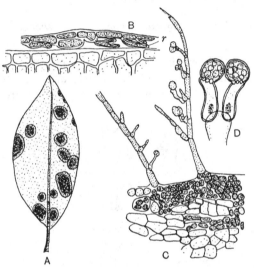

Fig. 48. *Cephaleuros*. A, leaf of *Magnolia* infected with *C. virescens*. B, trans-verse section of leaf of *Michelia fuscata* showing filaments and rhizoids (*r*) of *C. virescens*. C, transverse section of leaf of *Zizyphus* with *C. minimus* showing sporangial branches. D, sporangia of *C. mycoidea*. (A, after Smith; B, C, after Fritsch; D, after Oltmanns.)

hairs, whilst the parasitic species possess rhizoids which penetrate the cells of the host, although it has not been clearly established whether the host cells are killed before or after penetration. *Cepha-leuros virescens* forms the red rust of the tea plant which may cause much economic damage, but the attack is only serious when the tea tree is growing slowly, because during periods of rapid growth the alga is continually being shed by exfoliation of the outer tissues. The disease cannot be controlled by spraying with poisons, but the bushes can be made less susceptible to attack by treating the soil with potash. The genus is confined to the tropics.

*Trentepohliaceae: *Trentepohlia* (after J. F. Trentepohl) (*Chroolepus*). Fig. 49.

The species grow as epiphytes or on stones in damp tropical and subtropical regions, but they will also grow under temperate

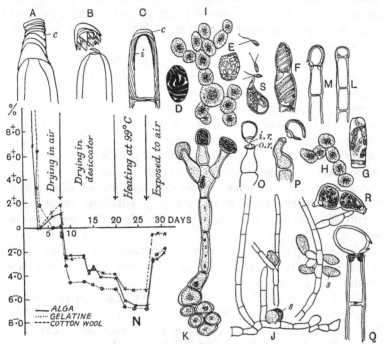

Fig. 49. *Trentepohlia.* A, B, *T. montis-tabulae* with pectin caps. C, *T. montis-tabulae*, cell structure. *c* = cap, *i* = innermost layer of cell wall. D–F, types of chloroplast. G, chloroplast in *T. Iolithus*. H, I, *T. umbrina*, fragmentation of prostrate system. J, threads of *T. aurea* bearing sporangia (*s*). K, *T. umbrina*, sporangia. L, M, two stages in the development of the "funnel" sporangium in *T. annulata*. N, graph showing decreasing water contents of *Trentepohlia*, gelatin and cotton-wool on drying. O, P, *T. umbrina*, detachment of stalked sporangium. *i.r.* = inner, *o.r.* = outer thickening of sporangial septum. Q, mature "funnel" sporangium, *T. annulata*. R, S, gametangia of *T. umbrina*. (A–G, J, L, M, O–Q, after Fritsch; H, I, K, R, S, after Oltmanns; N, after Howland.)

conditions if there is an adequate supply of moisture. The threads have a characteristic orange-red colouring due to the presence of β-carotin which is said to be a food reserve accumulated during periods of slow growth, but if this is so it would be expected that it should accumulate under favourable conditions of rapid growth and disappear under unfavourable conditions when growth is slow. This is a feature of its metabolism that would seem to require further investigation. The cells contain chloroplasts that are discoid or band-shaped and devoid of pyrenoids. Usually both prostrate and erect threads are present, though the latter are reduced in some species. Growth is apical, and the terminal cells often bear a pectose cap or series of caps which are periodically shed and replaced by new ones. The origin of the cap is not properly understood but it is thought to be due to a secretion, whilst its function may be either to reduce transpiration or else to act as a means of protection: alternatively, it may simply be a means of removing waste material. The cellulose walls are frequently thickened by parallel or divergent stratifications, whilst each septum between the cells may also have a single large pit which is penetrated by a protoplasmic strand. The cells are uninucleate when young and multinucleate when old, but the presence of the pigment makes the nuclei extremely difficult to distinguish. Vegetative reproduction is through fragmentation, whilst other means of reproduction are to be found in three different types of sporangia:

(a) *Sessile* sporangia that never become detached. These consist of enlarged cells which develop in almost any position and they produce biflagellate swarmers that may be isogametes.

(b) *Stalked* terminal or lateral sporangia that are cut off from an enlarged support cell which may give rise to several such bodies. The apical portion swells out to form the sporangium and cuts off a stalk cell underneath that frequently becomes bent. The dividing septum possesses two ring-shaped cellulose thickenings which may be connected with the detachment of the sporangium when it is mature. The detached sporangium is blown away and germinates under favourable conditions to bi- or quadriflagellate swarmers.

(c) *Funnel*-shaped sporangia which are cut off at the apex of a cylindrical cell, the outer wall splitting later at the septum, thus liberating the sporangium, the subsequent fate of which is not definitely known. The sessile and stalked sporangia may occur on the same plant or else on separate plants. There has been no cytological work to show whether there is any alternation of generations and such an investigation would be highly desirable. In one species, on the other hand, reproduction is wholly by means of aplanospores.

Howland (1929) has investigated the physiology of the commonest species, *T. aurea*, in some detail and he found that

(*a*) drought increases the resistance to plasmolysis;

(*b*) if the threads are dried first and then heated together with cotton-wool and gelatine, the results suggest that the threads hold water in a manner similar to that of cotton-wool, but that the loss of water on heating is comparable to that experienced by a colloid or gel under the same circumstances (cf. fig. 49);

(*c*) in damp, warm weather only small cells are formed because cell division is relatively rapid;

(*d*) the threads can survive desiccation for at least six months;

(*e*) plasmolysis could only be produced in some of the cells by a 25 % solution of sea salt.

In many respects, e.g. the heterotrichous nature of the thallus, the different types of sporangia and the orange pigment, this alga is strongly suggestive of the more primitive brown algae. This feature, however, is discussed more fully in a later chapter (cf. p. 255).

*COLEOCHAETACEAE: *Coleochaete* (*coleo*, sheath; *chaete*, hair). Fig. 50.

Most of the species are fresh-water epiphytes attached to the host by small outgrowths from the basal walls, but there is one species that is endophytic in *Nitella*, one of the Charales (p. 108). Some of the species are truly heterotrichous whilst others only possess the prostrate basal portion, which is either composed of loosely branched threads or else is a compact disk. The growth of the erect filaments is by means of the apical cell whilst the basal cushion possesses a marginal meristem. Each cell contains one chloroplast with one or two pyrenoids, and although a characteristic sheathed bristle arises from each cell nevertheless in the old plants these may be broken off. These bristles develop above a pore in the cell wall through which the protoplast extrudes, whilst at the same time a membrane is secreted over the protruding bare protoplast. Asexual reproduction takes place in spring and early summer by means of biflagellate zoospores which have no eye-spot and are produced singly. After a motile phase the zoospore settles down and divides either (*a*) horizontally, when the upper segment develops into a hair and the lower forms the embryo disk, or

(b) vertically, when each segment grows out laterally; in either case it will be noted that hair formation takes place at a very early stage.

Sexual reproduction is by means of a specialized oogamy, some of the species being dioecious and the remainder monoecious. The female organs, or *carpogonia*, are borne on short lateral branches and subsequently undergo displacement. Each carpogonium possesses a short neck or trichogyne (the long neck of *Coleochaete scutata* being an exception) the top of which bursts when the carpogonium is mature. In the disk forms the carpogonia originate as terminal

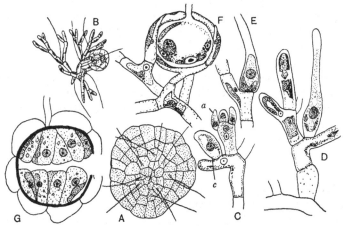

Fig. 50. *Coleochaete.* A, *C. scutata*, thallus with hairs (× 150). B, *C. pulvinata* with spermocarp (× 45). C, *C. pulvinata* with antheridia (*a*) and young carpogonium (*c*). D, *C. pulvinata*, almost mature carpogonium. E, *C. pulvinata*, fertilized carpogonium. F, *C. pulvinata*, formation of envelope around fertilized carpogonium. G, *C. pulvinata*, mature spermocarp with carpospores. (A, B, after Smith; C–G, after Fritsch.)

bodies on the outside of the disk, but as the neighbouring cells continue growth they eventually become surrounded and appear to be in the older part of the thallus. The antheridia develop in clusters at the end of branches (*C. pulvinata*) or from prostrate cells. They finally appear as small outgrowths cut off from a mother cell with stages in their development that are strongly reminiscent of the Rhodophyceae (cf. p. 252). Each antheridium produces one biflagellate colourless antherozoid which has been contrasted with the non-motile rhodophycean spermatium.

After fertilization the neck of the carpogonium is cut off and the basal part enlarges; branches arise from the underlying cells and eventually surround the oospore where they form a red or reddish brown wall, though in the disk forms this wall is only formed on the side away from the substrate. At the same time the enclosed oospore develops a thick brown wall and the cells of the outer envelope then die. The oospore, or *spermocarp*, hibernates until spring when it becomes green and divides into sixteen or thirty-two cells, and these, when the wall bursts, each give rise to a single swarmer which must be regarded as a zoospore. Meiosis takes place at the segmentation of the zygote so that there is only the haploid generation. On the other hand, some observers have recorded the development of dwarf asexual plants before the reappearance of new sexual ones, but this is a phase of the life history that demands reinvestigation, for if it is correct it may mean that there is an alternation of two unlike generations, an unusual phenomenon in the Chlorophyceae. Under certain conditions the cells will also produce aplanospores. The relation of this genus, with its advanced oogamy, to the other green algae is by no means clear, and although in many of its features the sexual reproduction is akin to that of the Rhodophyceae, it is commonly regarded as parallel evolution rather than as indicating a more direct relationship (cf. p. 256).

REFERENCES

Trentepohlia. BRAND, F. (1910). *Ber. dtsch. bot. Ges.* **28**, 83.
Trentepohlia. HOWLAND, L. J. (1929). *Ann. Bot., Lond.*, **43**, 173.
Stigeoclonium. REICH, K. (1926). *Arch. Protistenk.* **53**, 435.
Draparnaldia. USPENSKAJA, W. J. (1929–30). *Z. Bot.* **22**, 337.
General. VISHER, W. (1933). *Beih. bot. Zbl.* **51**, 1.
Coleochaete. WESLEY, O. C. (1928). *Bot. Gaz.* **86**, 1.

SIPHONOCLADIALES

Until 1935 this represented a well-established order, but in that year Fritsch placed most of the genera in the Siphonales but retained the Cladophoraceae as a separate order, the Cladophorales, with affinities to the Ulotrichales. More recently Feldmann (1938), in a survey of the group, has returned to the earlier idea of a relationship with the Siphonales via *Valonia* and *Halicystis*, though he also suggests relationships with *Chaetophora* and *Ulothrix*. Whatever the relations may be, the present order is clearly demarcated from the other groups and any affinities would seem to be somewhat distant.

*CLADOPHORACEAE: *Cladophora* (*clado*, branch; *phora*, bearing). Figs. 51, 52.

This is a widespread genus that occurs in both fresh and saline waters. The sessile forms are attached by means of branched septate rhizoids, but some of them (e.g. *C. fracta*) may become free-living later, whilst there is one complete section (*Aegagropila*) which is wholly free-living, the species existing as ball-like growths. The *Cladophora* thallus is composed of branched septate filaments, each cell usually being multinucleate, though cells with one nucleus have been recorded. The elongate reticulate chloroplasts, containing numerous pyrenoids, are arranged parietally with processes projecting into the central vacuole, but under some conditions they break up into fragments. There would not appear to be much present support for the old view that the chloroplast of each cell is a complex of numerous disk chloroplasts. The cell walls exhibit stratification as they are composed of three layers, an inner zone, a median pectic zone, and an outer zone which is said to be chitinous. There is very little production of mucilage, and this probably accounts for the dense epiphytic flora that is frequently found associated with species of this genus. The branches arise towards the upper end of a cell and later on are frequently pushed farther up, a process known as *evection*, thus giving the appearance of a dichotomy. All the cells are capable of growth and this is especially evident in cases of injury, but normally most of the plant growth is apical and in the section *Aegagropila* is wholly confined to the apex. At cell division the new septa arise from the outer layers and develop inwards, leaving in the process triangular-shaped spaces which later on may become filled with pectic substances or folded lamellae. Additional supporting rhizoids usually develop from the basal and subbasal cells of the lowest branches.

In the *Aegagropila* group the species can exist as (*a*) threads, (*b*) cushions and (*c*) balls. The destruction of the old threads in the centre of the ball results in a cavity which may become filled with water, gas or mud. In Lac Söro the water in April and May is sufficiently free of diatoms for light to penetrate to such an extent that photosynthesis increases and so much gas collects in the centre of these

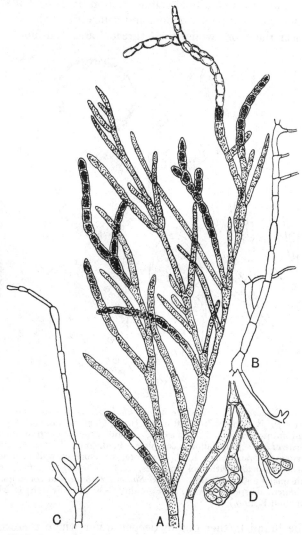

Fig. 51. *Cladophora.* A, plant with sporangia. B, shoot of *Aegagropila holsatica* bearing rhizoids. C, stolon of *Ae. holsatica.* D, rhizoids of *Spongomorpha vernalis* developing storage cells at the apices. (A, after Oltmanns; B, C, after Acton; D, after Fritsch.)

balls that they float to the surface. Their characteristic shape is brought about by a continual rolling motion over the soil surface under the influence of wave action, and hence the "ball" forms are found near the shore whilst the "thread" and "cushion" forms

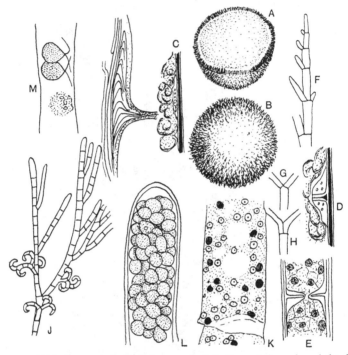

Fig. 52. *Cladophora.* A, ball of *Aegagropila holsatica* cut through and the dirt removed (× ⅔). B, same before cutting (× ⅔). C, *C. glomerata*, commencement of septum formation. D, *C. glomerata*, second stage in septum formation. E, *C. glomeratum*, septum almost complete. F, diagram illustrating evection. G, H, types of branching. I, *C. glomerata*, structure of wall at a septum. J, *Spongomorpha coalita* with hook branches. K, *C. callicoma*, structure of chloroplast with nuclei and pyrenoids. L, *Ae. Sauteri*, zoospores in zoosporangium. M, *Ae. Sauteri*, zoospores. (A, B, after Acton; C–K, after Fritsch; L, M, after Nishimura and Kanno.)

are to be found farther out in deeper water where there is less motion. The harder the floor the more regular is the shape of the balls, but even so the ball structure would also appear to be inherent in the alga because "balls" have been kept in a laboratory for eight years without losing their shape. The following types of branches have been recognized in the *Aegagropila* forms: (*a*) rhizoids, (*b*) cirrhoids, both these and the rhizoids being neutral or non-reproductive branches, (*c*) stolons or vegetative reproductive branches. Many of the species of *Cladophora* are perennial, and in the section *Spongomorpha* the rhizoids form a basal expanse from which new threads may arise each year. In some of the fresh-water species certain cells may become swollen to form akinetes in which the walls are thickened and food is stored.

In the section *Aegagropila* most of the species reproduce vege-tatively, but biflagellate swarmers have been reported for one species, *Ae. Sauteri*, and these are interesting in that they may germinate whilst still within the sporangium (fig. 52 L, M). Asexual reproduction in the other species, excluding the section *Aegagropila*, is by means of quadriflagellate zoospores (bi-flagellate in two species) which escape through a small pore in the cell wall. Biflagellate isogametes are the means of sexual repro-duction, all the species so far investigated being dioecious. The zygote develops at once without a resting period. In a number of species alternation of two morphologically identical haploid and diploid generations has now been established with meiosis taking place at zoospore formation. In one or two cases, e.g. *Cladophora flavescens*, the zoospores sometimes fuse, and this irregular be-haviour is very comparable to similar phenomena found in the more primitive brown algae (cf. p. 138).

In a few species there is an odd or heterochromosome, and in a cell the number of zoospores with the odd chromosome are equal to the number lacking it. Haploid plants of *C. Suhriana* have six or seven chromosomes, whilst in *C. repens* the cells contain either four or five. In a fresh-water species, *C. glomerata*, a wholly different type of life cycle is known, and this difference may perhaps be compared with the various cycles found for *Ectocarpus siliculosus* under different conditions (cf. p. 135). Gametes and zoospores are both formed on diploid plants and meiosis takes place at gamete formation so that there is no haploid generation. Whilst zoospore formation takes place all the year round gametes only appear in the spring, but the reason for this seasonal restriction is not under-stood. Parthenogenetic development of gametes has also been recorded in a number of species. Of the species so far investigated the chromosomes appear to be present in multiples of 4, and this probably indicates polyploidy. The following diploid chromosome numbers have been recorded: *C. repens* 8 + 1, *C. Suhriana* 12 + 1, *C. flavescens* 24, *C. flaccida* 24, *C. pellucida* 32, *C. glomerata* 64.

CLADOPHORACEAE: *Rhizoclonium* (*rhizo*, root; *clonium*, branch). Fig. 53.

This genus is either marine, brackish or fresh water, several marine species being found in great quantities on sand or mud flats. The uniseriate filaments are simple or else possess short septate or non-septate colourless rhizoidal branches. The threads are smaller in diameter than those of the preceding genera, and the

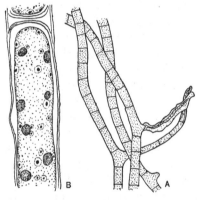

Fig. 53. *Rhizoclonium*. A, part of filaments of *R. riparium* (× 90). B, cell of *R. hieroglyphicum* to show structure of chloroplast. (A, after Taylor; B, after Fritsch.)

number of nuclei per cell are also less, usually one to four, although in the stouter species there may be as many as twenty-four. It has been found that the number of nuclei contained depends on the cubical contents of the cell, a feature of size and form that is analogous to the phenomenon found in the higher plants. The number of epiphytes may also influence the quantity of the nuclei. The plants are attached at first but become free-living later, and in this state some of the larger species are scarcely distinguishable from small species of the related genus *Chaetomorpha*. Vegetative reproduction is by means of biflagellate zoospores which in some species are said to have unequal flagellae. Anisogamy similar to that of the related genus *Urospora* has been recorded for *Rhizo-clonium lubricum*. *Urospora* is of interest because the zygote first produces a *Codiolum* stage (so called after the alga it resembles), which is considered to be diploid, and this gives rise to zoospores from which the normal filaments develop, so that if this interpretation of the life history is correct we have here another rare example of alternation of morphologically dissimilar generations in the Chlorophyceae.

*VALONIACEAE: *Valonia* (after the Valoni, an Italian race), "Sea-Bottle". Fig. 54.

In this genus, which is restricted to warm waters, the young coenocyte consists of one large vesicle whilst the old one becomes

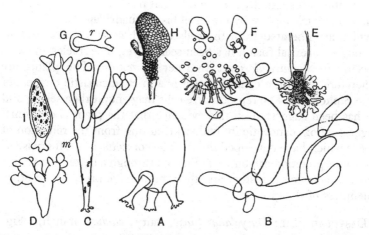

Fig. 54. *Valonia*. A, young plant of *V. ventricosa*. B, young plant of *V. utricularis* (× 1·4). C, adult plant of same. *m* = marginal cell. D, plant of *V. macrophysa* (× 0·8). E, rhizoid of *V. utricularis*. F, rhizoids from marginal cells at base of vesicle of *V. ventricosa*. G, single marginal cell and rhizoid (*r*) of *V. ventricosa*. H, *V. utricularis* fruiting. I, *V. utricularis*, germinating swarmer. (B, D, after Taylor; rest after Fritsch.)

divided up into a number of multinucleate segments. It has been suggested (cf. p. 280) that it should really be regarded as a coeno-cytic wall enclosing a fluid, but this interpretation leads to diffi-culties. In some respects, therefore, the genus provides a link with the Siphonales. The macroscopic club-shaped vesicle is attached to the substrate by rhizoids of various types. There is a lobed chloroplast that congregates with the cytoplasm at certain points in the older plants and then each group is cut off by a membrane, thus producing a number of marginal cells. This septation is regarded as a primitive character that is slowly being lost because in the more advanced Siphonales it is restricted to the formation of the reproductive organs. The cells do not necessarily form a continuous layer and are frequently restricted to the basal region where they may develop rhizoids, whilst in other species they are nearer to the apex where they may give rise to proliferations. The lower cells can form short creeping branches, and as these bear more of the erect vesicles a tuft of plants is produced. One genus (*Siphonocladus*), classed either in the Valoniaceae or else in a separate group, resembles *Cladophora* very closely although the method of segmentation is essentially the same as that of *Valonia*. Reproduction in *Valonia* takes place by means of bi- or quadriflagellate swarmers which are liberated from the cells through several pores, and although no sexual fusion has been seen as yet, nevertheless meiosis occurs in *V. utricularis* at swarmer formation. The plants are therefore diploid, a condition that is also characteristic of most of the Siphonales. The reproductive cells may encyst themselves, and it has been suggested on this evidence that the plant is a colonial aggregate of coenocytic individuals resulting from the retention of cysts which have developed *in situ*. The correctness or otherwise of this interpretation can only be obtained through a better knowledge of its phylogenetic history and the reproductive processes of other members of the group.

*DASYCLADACEAE: *Dasycladus* (*dasy*, hairy; *cladus*, branch). Fig. 55.

The family Dasycladaceae is very ancient and was formerly much more widely spread since sixty fossil genera are known whilst there are only ten living to-day (cf. p. 269). *Dasycladus* forms dense growths, up to 5 cm. in height, in shallow waters where the plants are anchored by means of richly branched non-septate rhizoids. The central axis bears dense whorls of profusely branched laterals which are arranged alternately above each other. The branches arise in whorls of four immediately below the apex of the parent cell, to which they are united by narrow constrictions, and although the rest of the main axis is impregnated with lime throughout there is none at the constrictions. If the axis or a branch is decapitated a new apex is regenerated, whilst if a rhizoid is cut off and inverted it develops a normal apical cell. Short-stalked spherical gametangia arise at the apices of the major branches in the upper half of the plant and are cut off by a septum. The plants are essentially dioecious and produce isogametes that sometimes exhibit relative sexuality.

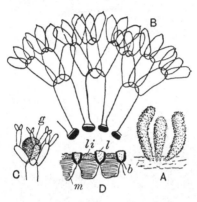

Fig. 55. *Dasycladus clavaeformis*. A, plants. B, assimilatory filaments showing mode of branching. C, gametangium (*g*). D, thickenings at base of assimilatory filaments. *b* = point of origin of branch, *l* = base of lateral, *li* = calcified wall, *m* = thickened base of wall. (After Fritsch.)

DASYCLADACEAE: *Neomeris* (*neo*, new; *meris*, part). Fig. 56.

This is a calcareous tropical genus which has been in existence from the Cretaceous era. The much calcified adult plants have the appearance of small worm-like masses with an apical tuft of hairs, whilst very young plants consist of an erect *Vaucheria*-like filament with a tuft of dichotomously branched filaments at the apex. In the adult plant the ultimate branches terminate in long deciduous hairs, whilst the apices of the next lower order of branches dilate and become pressed together, thus producing a compact surface with a pseudo-parenchymatous appearance (cf. fig. 56 E). Calcium carbonate is deposited wherever there is a mucilage layer and an aggregation of the chloroplasts, but apparently both these conditions must be fulfilled before lime can be laid down. The principal interest of this form lies in its morphological resemblance to certain fossil genera (cf. p. 271).

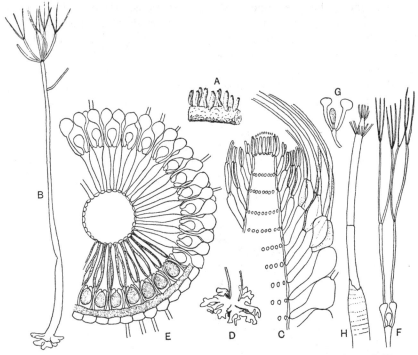

Fig. 56. *Neomeris*. A, plants of *N. annulata* ($\times \frac{1}{2}$). B, young plant of *N. dumetosa* ($\times \frac{1}{4}$). C, longitudinal section through apex of *N. dumetosa* ($\times \frac{1}{3}$). D, rhizoid in *N. dumetosa* ($\times \frac{1}{4}$). E, transverse section of thallus of *N. dumetosa* in middle of calcified area ($\times \frac{1}{4}$). F, *N. dumetosa*, assimilating filaments with sporangium ($\times \frac{1}{3}$). G, *N. annulata*, sporangium ($\times 33$). H, regeneration of an injured axis ($\times \frac{1}{4}$). (A, G, after Taylor; rest after Church.)

*DASYCLADACEAE: *Acetabularia* (*acetabula*, little cup; *aria*, derived from). Fig. 57.

This is a lime-encrusted genus which is confined to warm waters, extending up as far as the Mediterranean in the northern hemisphere. The plants consist of an erect elongate axis bearing one or more whorls of branched sterile laterals with a single fertile whorl at the apex. The sterile whorl or whorls are frequently shed in the adult plant leaving a mark or annulus on the stem to show where they were formerly attached. The fertile whorl is composed of a series of long sac-like sporangia which are commonly fused, though they are sometimes separate: these are borne on short basal segments which are morphologically equivalent to the primary branches. The basal segments also bear on their upper surface small projections, with or without hairs, which form the *corona*, whilst in one section of the genus there is also an inferior corona on

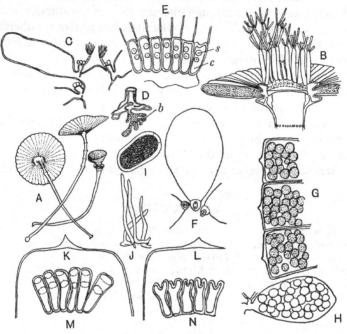

Fig. 57. *Acetabularia*. A, plant of *A. crenulata* (×0·8). B, apex of *A. mediterranea* showing corona. C, apex of *A. Moebii* showing two superposed fertile whorls. D, *A. mediterranea*, attachment rhizoid and perennating vesicle (*b*). E, *A. crenulata*, cells near centre of thallus, showing superior corona (*c*) and leaf scars (*s*). F, *A. pusilla*, vegetative ray segment (×44). G, fertile lobes of *A. Schenkii* with cysts (×44). H, cysts in *A. pusilla* in a single lobe of the umbrella (×37). I, single cyst of *A. mediterranea*. J, young plant in first year. K, L, *A. crenulata*, apices of ray segments (×37). M, *A. crenulata*, superior corona (×37). N, *A. crenulata*, inferior corona (×37). (A, F–H, K–N, after Taylor; B–E, I, J, after Fritsch.)

the lower surface. In *A. mediterranea* two or three years elapse before the plant attains to maturity. In the first year the branched holdfast produces an upright umbilical thread, together with a thin-walled lobed outgrowth that penetrates the substrate in order to function as the perennating organ. The aerial part dies, and in

84 CHLOROPHYCEAE

the next year or years a new cylinder arises that bears one or more
sterile whorls of branches, until in the third or even a later year, a
shoot develops which produces one deciduous sterile whorl and a
single fertile whorl or umbrella. Each sac-like sporangium, or
umbrella lobe, gives rise to a number of multinucleate cysts which
are eventually set free through disintegration of the anterior end of
the sporangium. In the spring biflagellate isogametes are liberated
from these cysts and fuse in pairs, or else develop parthenogenetically. In *A. Wettsteinii* meiosis occurs at gametogenesis and the
adult plants are therefore diploid. According to Hammerling
the immature plant contains only one nucleus, which is to be
found in one of the rhizoids, and at cyst formation this divides, the
daughter nuclei being carried into the sporangia. The resulting
cysts in the umbrella lobes are uninucleate, but as the single
nuclear condition is in direct contrast to the reports of other
workers it would seem that further cytological study is desirable.

REFERENCES

Cladophora. ACTON, E. (1916). *New Phytol.* **15**, 1.
Neomeris. CHURCH, A. H. (1895). *Ann. Bot., Lond.*, **9**, 581.
General. FELDMANN, J. (1938). *Rev. Gen. Bot.* **50**, 571.
Acetabularia. HAMMERLING, J. (1931, 1932). *Biol. Zbl.* **51**, 663; **52**, 42.
Cladophora. LIST, H. (1930). *Arch. Protistenk.* **72**, 453.
Neomeris. SVEDELIUS, N. (1923). *Svensk bot. Tidskr.* **17**, 449.

SIPHONALES

This group is characterized primarily by possession of a coenocytic structure in which true septa are rare or absent, the coenocyte
normally having a cytoplasmic lining surrounding a central vacuole
and containing numerous disk-shaped chloroplasts. Cellulose is
largely replaced by callose as the principal component of the walls.
The group may be polyphyletic in origin, and the fact that it
reaches its maximum development in warm waters may be significant, not only in respect of the phylogeny of the group itself, but
also in considering the evolution of the Chlorophyceae as a whole.
Most of the genera possess the power of regeneration to a marked
degree, but this can perhaps be regarded as a primitive character
that has persisted throughout the course of time.

PROTOSIPHONACEAE: *Protosiphon* (*proto*, first; *siphon*, tube). Fig. 58.

The single species common in north Europe grows in damp mud at the edges of ponds, but a variety is also known from the desert silt of Egypt which will tolerate temperatures up to 91° C. and salt concentrations of at least 1 %. The green aerial portion is more or less spherical, up to 100 μ. in diameter, grading into a colourless rhizoidal portion that is occasionally branched. The chloroplast, which contains numerous pyrenoids and nuclei, is an anomaly in

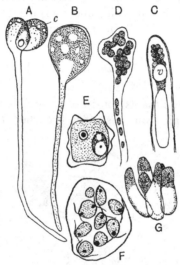

Fig. 58. *Protosiphon botryoides*. A, B, plants, one showing budding. c = chloroplast. C, swarmer formation. v = vacuole. D, cyst formation. E, zygote ($\times$ 1666). F, germination of zygote to form zoospores ($\times$ 1026). G, group of plants grown in a nutrient solution. (A–D, G, after Fritsch; E, F, after Bold.)

the group because of its reticulate character. In very dry places the rhizoid may be abbreviated to such an extent that the plant looks like a *Chlorococcum*, whilst in cultures where nutrient conditions are favourable one may obtain branched thread-like growths. The shape of the thallus is determined by the incidence of the light, unilateral light producing asymmetrical aerial portions, whilst exposure to bright light and low moisture may also cause an old thallus to turn brick red. During times of drought resting spores or cysts are formed which, when conditions become favourable once more, either germinate directly or else produce zoospores, germination in the desert forms occurring at temperatures between 12° and 35° C. Vegetative reproduction takes place by means of lateral budding, but when submerged the plants also produce naked biflagellate swarmers which usually act as isogametes, though they are also capable of parthenogenetic development. *Protosiphon botryoides* is monoecious whilst the desert variety is dioecious, and this fact alone would seem sufficient justification for regarding the latter as a distinct species. The zygote either germinates immediately to give a new plant or else may remain dormant for some time. The plant is probably haploid, and morphologically is of great interest in indicating how the more advanced Siphonales may have arisen.

*HALICYSTACEAE: *Halicystis* (*hali*, salt; *cystis*, bladder) and *Derbesia* (after A. Derbes). Fig. 59.

The gametophytic plants consist of an oval vesicle, up to 3 cm. in diameter, arising from a slender branched tuberous rhizoid

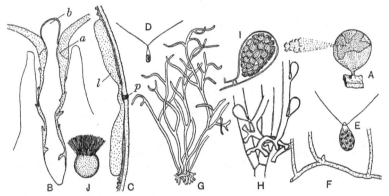

Fig. 59. *Halicystis ovalis* (and *Derbesia marina*). A, plant of *Halicystis* liberating gametes. B, rooting portion of *Halicystis* showing old rhizome and line of abscission (*a*) and new vesicle (*b*). C, gathering of protoplasm to form gametes. *l* = lining cytoplasm, *p* = pore of dehiscence. D, male gamete (× 600). E, female gamete (× 600). F, protonemal germling of *Halicystis*. G, *Derbesia* plant. H, *Derbesia*, with zoosporangia, growing on *Cladophora*. I, *Derbesia*, zoosporangium. J, *Derbesia*, zoospore. (A–C, F–J, after Fritsch; D, E, after Kuckuck.)

embedded in calcareous *Lithothamnia* (cf. p. 317) growing at or below low-tide mark. There are only two species of *Halicystis*, one of which possesses pyrenoids whilst the other does not, though both contain numerous nuclei in the peripheral cytoplasm. There does not appear to be any cellulose in the material composing the cell wall. Swarmers develop in the cytoplasm at the apex of the vesicle in an area which becomes cut off by a thin cytoplasmic membrane, the area thus cut off representing a gametangium. Macro- and microgametes are formed and forcibly discharged in the early hours of the morning through one or more pores. There are several crops of these swarmers produced by successive migrations of cytoplasm into the apical areas at bi-weekly intervals coincident with the spring tidal cycles. Fertilization occurs in the water, and the zygote in *H. ovalis* germinates into a branched protonemal thread that in 3 months has developed into a typical *Derbesia* plant with the erect aerial filaments arising from the basal rhizoidal portion.

It has been demonstrated only quite recently that both *Halicystis ovalis* and *Derbesia marina* are simply two stages in the life cycle of one alga, but in addition to the evidence from cultures the two species have the same geographical distribution. The mature *Derbesia* threads produce zoospores that germinate into prostrate filaments, and these later give rise to slender branched rhizoids which, after eight months, produce the characteristic *Halicystis* bladder. Some weeks after its development the bladder becomes fertile and so the cycle starts once more. Although the cytology of the two plants has not yet been worked out the *Derbesia* generation is presumably diploid and the *Halicystis* haploid. It also remains to be ascertained whether the other species of *Halicystis* has a similar life cycle. Growth of the *Halicystis* vesicles is very slow and they become shed at the end of the growing season by abscission, new vesicles arising later from the perennating rhizoid, and in this manner regeneration may go on for several years. The genus, formerly regarded as a connecting link between *Protosiphon* and members of the Valoniaceae, must now be removed into a separate family because of this remarkable life history. This new family must also include *Derbesia* in the same way that *Aglaozonia* is now included in *Cutleria* (cf. p. 156).

PHYLLOSIPHONACEAE: *Phyllosiphon* (*phyllo*, leaf; *siphon*, tube). Fig. 60.

This is an endophytic alga that occurs in the leaves and petioles of the Araceae, most of the species being confined to the tropics,

CHLOROPHYCEAE

although one is found in Europe, including Great Britain. The thallus is composed of richly branched threads ramifying in the intercellular spaces of the host. As a result of the presence of the endophyte the chloroplasts of the host cells do not develop and yellow-green patches occur on the leaf, whilst at the same time the adjacent cells may be stimulated to active division resulting in gall formation, but later on the affected cells die. Reproduction takes place by means of oval aplanospores.

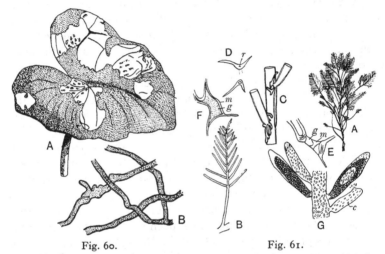

Fig. 60.   Fig. 61.

Fig. 60. *Phyllosiphon Arisari.* A, leaf of *Arisarum vulgare* with whitened patches due to attack of alga. B, portion of thallus (× 66). (A, after Fritsch; B, after Smith.)

Fig. 61. *Bryopsis.* A, plant of *B. plumosa* (× 0·6). B, portion of same enlarged (× 7). C, *B. corticulans*, rhizoid formation from lower branches. D–F, stages in septum formation at base of gametangium. *g* = gelatinized material, *m* = membrane, *r* = ring of thickening initiating septum. G, *B. plumosa*, female gametangia. *c* = chloroplast. (A, B, after Taylor; C–G, after Fritsch.)

*CAULERPACEAE: *Bryopsis* (*bryo*, moss; *opsis*, an appearance). Fig. 61.

Most of the species of this genus are restricted to warmer seas, though at least two, of which *B. plumosa* is the commoner, occur in colder waters. The principal axis, which is often naked in its lower part, arises from an inconspicuous, filamentous, branched rhizome that creeps along the substrate and is attached to it by means of rhizoids. In one species the bases of the lower branches develop additional rhizoids that grow down and form a sheathing pseudo-cortex. The bi- or tripinnate fronds usually have the branching confined to one plane, the branches being constricted at the point of origin, whilst the cell membrane is also thickened at such places. The cytoplasm in the main axis and branches frequently exhibits streaming movements. The function of the rhizome, especially in warmer waters, is probably that of a perennating organ, although vegetative multiplication can also occur through abstriction of the pinnae, which then develop rhizoids at their lower end. The only other known method of reproduction is sexual. The plants are dioecious and produce anisogametes which develop in gametangia that are cut off from the parent thallus by means of septa. Both types of gamete are biflagellate, but the microgametes differ from the macrogametes in that they lack pyrenoids. The gametes are liberated through gelatinization of the apex of the gametangium, and after fusion has taken place the zygote germinates at once into a new plant. The plants are diploid because meiosis takes place at gamete formation; there is therefore no haploid generation. The plants can behave like *Vaucheria* (cf. p. 95) in their response to certain environmental conditions; thus, gamete formation is hastened by transference of the plants from light to dark or by changing the concentration of the nutrient solution. Inversion of the thallus takes place under conditions of dull light or when it is planted upside down, and under these circumstances the apices of the pinnae develop rhizoids. This exhibition of polarity indicates clearly that the thallus is differentiated internally, but it is still a matter for speculation as to how such differentiation can occur in an organism which is to all intents and purposes one unit.

*CAULERPACEAE: *Caulerpa* (*caul*, stem; *erpa*, creep). Fig. 62.

Most of the species frequent the quiet shallow waters of the tropics where they are often rooted in sand or mud, but two have migrated far enough north to become denizens of the Mediterranean. The prostrate rhizome is attached by means of colourless rhizoids and gives rise to numerous erect, upright, assimilatory shoots with apical growth, the form and arrangement of which may vary very considerably (fig. 62 A–F). Radial branching is regarded as primitive, whilst the more evolved forms of quieter waters possess a bilateral branching system. The genus has been divided by Börgesen into three groups:

(*a*) The species of this group, which grow where there is much mud, possess rhizomes that are vertical or oblique, thus enabling them to reach the surface even when covered successively by mud (e.g. *C. verticillata*).

(*b*) The rhizome in these species first branches at some distance from its point of origin and it possesses a pointed apex which aids in boring through sand or mud (e.g. *C. cupressoides*).

(*c*) The rhizome is richly branched immediately from its point of origin and the various species are principally to be found attached to rocks and coral reefs (e.g. *C. racemosa*).

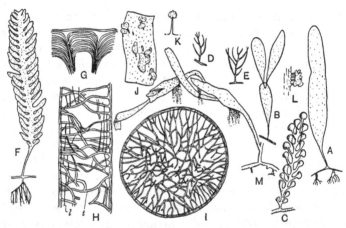

Fig. 62. *Caulerpa*. A, B, *C. prolifera* (× ½). C, *C. racemosa* f. *macrophysa* (× ½). D, E, *C. sertularioides*, side branches (× ½). F, *C. crassifolia* f. *mexicana* (× ½). G, structure of wall and two skeletal strands. H, longitudinal section of aerial portion showing longitudinal (*l*) and transverse (*t*) support strands. I, transverse section of rhizome with skeletal strands. J, K, L, *C. prolifera*, reproductive papillae (× 5). M, *C. prolifera* with gametes being liberated. (A–F, after Taylor; G–I, after Fritsch; J–M, after Dostál.)

It has also been shown that the form of the thallus in some of the species is largely dependent upon the conditions of the habitat, a feature particularly well illustrated by the plastic *C. cupressoides* and *C. racemosa*:

(i) In exposed situations the plants are small and stoutly built.

(ii) In more sheltered habitats the shoots are longer and more branched.

(iii) In deep water the plants are very large with richly branched flabellate shoots.

There is no septation, but the coenocyte is traversed instead by numerous cylindrical skeletal strands, or *trabeculae*, arranged perpendicularly to the surface and which are most highly developed in the rhizomes. They arise from rows of structures termed *microsomes*, and are at first either free in the interior of the coenocyte or else connected with the wall, although in the adult state they are always fused to the walls. The function of the trabeculae, which increase in thickness at the same time as the walls by successive deposition of callose, is extremely problematical and may be

(*a*) mechanical: in this case they would presumably provide resistance to high turgor pressures, although the presence of high osmotic pressures in the cells has yet to be proved;

(*b*) to enlarge the protoplasmic surface;

(*c*) concerned with diffusion, because movement of mineral salts is more rapid through these strands than through the cytoplasm;

(*d*) lost or without any function.

In addition to the trabeculae there are also internal peg-like projections. Vegetative reproduction occurs through the dying away of portions of the old rhizome thus leaving a number of separate plants. The swarmers or gametes are formed in the aerial portions and are liberated through special papillae that develop on the frond. The sexual reproductive fronds have a variegated appearance caused by the massing of the biflagellate gametes at the different points, the swarmers in some species being separable into micro- and macrogametes. In certain species the whole plant can produce swarmers, whilst in others the reproductive area is limited, and in such cases the morphological identity and differentiation of the frond becomes of great interest. The thallus can be regarded as composed of a number of individual cells which only become evident at gametogenesis. Fusion between the swarmers has been observed in *C. racemosa*, and it is probable that in all the species the motile bodies are functional gametes and that the adult plants are diploid. The genus has been much employed in experiments on polarity because the structure of the thallus renders it extremely suitable.

CODIACEAE: *Codium* (fleece). Fig. 63.

This is a widely distributed, non-calcareous genus with several species living in the colder oceans. The erect and fleecy thallus, which is anchored either by a basal disk or else by rhizoids, varies greatly in form and appears as branched worm-like threads, flat

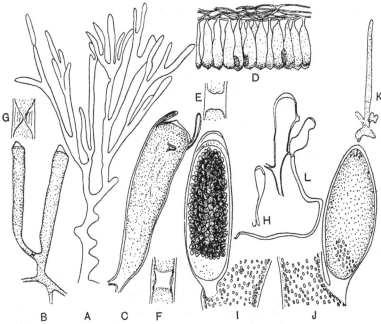

Fig. 63. *Codium*. A, plant of *C. tomentosum*. B, *C. fragile*, utricles. C, *C. tomentosum*, single utricle with hairs. D, *C. tomentosum*, portion of thallus with medulla and cortical utricles. E–G, stages in formation of constriction at base of utricle. H, propagule of *C. isthmocladum*. I, *C. tomentosum*, female gametangium. J, *C. tomentosum*, male gametangium. K, *C. tomentosum*, juvenile thread. L, *C. isthmocladum*, utricle with propagule. (A, after Taylor; B, C, E–G, J, after Tilden; D, H, K, L, after Fritsch; I, after Oltmanns.)

cushions, or as large round balls. In *C. tomentosum* there is a central medulla of narrow forked threads and a peripheral cortex of club-shaped vesicles which are the swollen apices of the forked threads. Deciduous hairs may develop on the vesicles and scars are to be seen marking their point of attachment, whilst annular thickenings occur at the base of each vesicle and at the bases of the lateral branches, although a fine pore is left for intercommunication.

The width of these pores in the case of *C. Bursa* is said to vary with the season. Detachable propagules develop on the vesicles and form a method of vegetative reproduction, whilst sexual reproduction is by means of gametes, which are produced in ovoid gametangia that arise from the vesicles as lateral outgrowths, each being cut off by a septum. The plants are anisogamous, the macrogametes being formed in green and the microgametes in yellow gametangia. Some of the species are dioecious whilst others are monoecious, and in two of the latter the male and female gametangia are borne on the same utricles. The gametes fuse or else develop parthenogenetically, but in either case a single thread-like protonema develops which has a lobed basal portion, and it is from this that the adult develops through the growth of numerous ramifications of the one primary filament. Meiosis occurs at gametogenesis and the plants are therefore wholly diploid and comparable to *Fucus* (cf. p. 192). At gametogenesis some of the nuclei in the gametangia degenerate whilst the remainder divide twice.

CODIACEAE: *Halimeda* (*Halimeda*, daughter of Halimedon, King of the sea). Fig. 64.

The genus is known from Tertiary times onwards, and it has played a considerable part in the formation of coral reefs where the

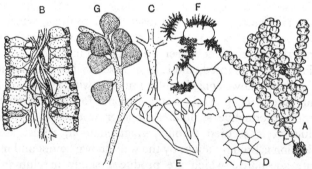

Fig. 64. *Halimeda.*   A, plant of *H. simulans* (× 33). B, *H. discoidea*, longitudinal section showing structure (× 20). C, central filament: two fuse and subsequently divide into three (× 20). D, cuticle of *H. opuntia* (× 132·5). E, *H. scabra*, termination of filaments (× 100). F, fruiting plant. G, sporangia. (A, D, E, after Taylor; B, C, after Howe; F, G, after Oltmanns.)

species are very abundant. The plants are borne on a short basal stalk that arises from a prostrate system of creeping rhizoids. The branched aerial thallus is composed of flat, cordate or reniform segments which are strongly calcified on the outside, the segments being separated from each other by non-calcified constrictions. The segments are composed of interwoven threads with lateral branches that develop perpendicularly and produce a surface of hexagonal facets through fusion of the swollen ends. Sporangia develop at the ends of forked threads which vary greatly in their mode of branching: these threads, which are cut off from the parent thallus by basal plugs, arise from the surface of the segments or, more frequently, are confined to the edges. The sporangia produce biflagellate swarmers whose fate is not known although they are probably gametes.

VAUCHERIACEAE: *Vaucheria* (after J. P. Vaucher). Figs. 65, 66.

This genus differs in many of its characters from the other members of the Siphonales, and it should perhaps be removed into the Xanthophyceae. Whereas most of the Siphonales are tropical genera *Vaucheria* is essentially temperate, inhabiting well-aerated streams, soil or saline mud flats, and although some of the species (e.g. *V. Debaryana*) may be lime-encrusted it is never to quite the same extent as in the preceding genera. There is a colourless basal rhizoidal portion from which arise green, erect aerial filaments with apical growth and monopodial branching. The cell walls contain cellulose and pectins whilst the discoid chloroplasts, which lack pyrenoids, contain more than the normal amount of xanthophyll. Oil forms the principal food reserve, except that under constant illumination starch is formed, and it is in these biochemical characters that *Vaucheria* shows considerable similarity with members of the Xanthophyceae (cf. p. 113). Septa are only formed in connexion with the reproductive structures or after wounding. Vegetative reproduction is secured through fragmentation, whilst asexual reproduction is brought about by the well-known compound multi-flagellate zoospores, which are produced singly in club-shaped sporangia that are cut off from the ends of the erect aerial branches. The chloroplasts and nuclei congregate in the apex of a filament before the septum is laid down and the nuclei then arrange themselves peripherally. Finally, two equal flagellae develop opposite each nucleus and then the zoospore is ready for liberation, a process which is achieved by gelatinization of the sporangium tip. This compound structure must be regarded as representing a group of biflagellate zoospores which have failed to separate. The zoospore is motile for about 15 min., after which it comes to rest and germinates, the first thread often being more or less colourless. "Zoospore"

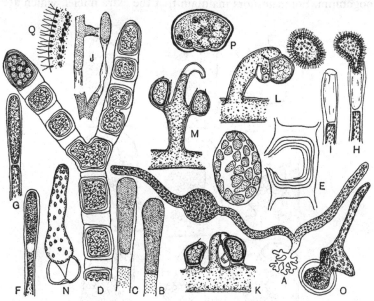

Fig. 65. *Vaucheria.* A, *V. sessilis*, germinating zoospore. B, *V. piloboloides*, developing aplanospore. C, *V. piloboloides*, escape of aplanospore. D, *V. geminata*, thread with cysts. E, escape of amoeboid protoblast from cyst. F–I, *V. repens*, development and escape of compound zoospore. J, regeneration and formation of septa in injured thalli. K, sex organs of *V. sessilis* (× 100). L, sex organs of *V. terrestris* (× 100). M, sex organs of *V. geminata* (× 100). N, *V. geminata*, germinating aplanospore. O, germinating zygote. P, zygote with four haploid nuclei. Q, portion of compound zoospore, much magnified. (A, D, E, N, O, after Oltmanns; B, C, F–J, Q, after Fritsch; K–M, after Hoppaugh; P, after Hanatschek.)

formation can often be induced by transferring the plants from light to darkness, or from a nutrient solution to distilled water.

Under dry conditions aplanospores may be formed at the ends of short laterals or terminal branches, whilst if exposed to greater desiccation the threads of the terrestrial forms become septate and rows of cysts are formed, thus giving the "*Gongrosira*" stage. When conditions become more favourable these cysts germinate either into new filaments or else into small amoeboid

masses which grow into new filaments. Sexual reproduction is
distinctly oogamous, the different species being either monoecious
or dioecious. The oogonia, which are sessile or stalked, are cut off
by a septum at a stage when there is only one nucleus left in the
oogonium. Some authors maintain that the extra nuclei, which are

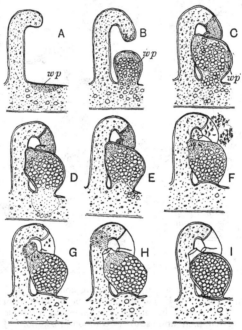

Fig. 66. *Vaucheria sessilis*. Stages in development and fertilization of oogonium.
April 1–6, 1930. (×195.) A, young antheridium and "wanderplasm" in place
from which oogonium will arise. B, young oogonium. C, oogonial beak formed;
"wanderplasm" retreating into thread; oil globules passing into oogonium;
antheridial wall forming. D, "wanderplasm" out of oogonium. E, basal wall of
oogonium forming. F, antherozoids emerging. G, oogonial membrane forming
at tip, some antherozoids in egg. H, cytoplasm extruded and rounded off;
fertilization occurring. I, ripe egg. *wp* = wanderplasm. (After Couch.)

potential gametes, degenerate, whilst others consider that the
surplus nuclei, enclosed in a mass of cytoplasm or "wanderplasm",
travel back into the main thread before the septum is laid down. It
is probable that in some species all the surplus nuclei pass out with
the "wanderplasm", whilst in other species some nuclei may be
left behind and degenerate later after the septum has been laid

down. The factors that determine the selection of the functional nucleus from among the number available offer a problem for future research. In the mature oogonium there is either a beak, the apex of which gelatinizes, or else several pores through which the antherozoids can enter the oogonium, fertilization taking place *in situ*.

The antheridia, which are usually stalked, commonly arise close to the oogonia, though in *V. sessilis* they develop just prior to oogonial formation. When the septum cutting off the antheridium is laid down the nuclei divide, and cytoplasm gathers around each daughter nucleus. The mature antheridium may be colourless or green, and it opens by one or more apertures near the apex, thus providing a means of escape for the pear-shaped antherozoids which bear two flagellae pointing in opposite directions. After fertilization the zygote develops a thick wall and remains dormant for some time before it germinates to give rise to a new filament. The latest evidence shows that reduction of the chromosome number takes place when the zygote germinates, thus indicating that the adult plant is haploid. This character is somewhat anomalous when contrasted with the diploid status of almost all the other Siphonales, with the exception of the primitive *Protosiphon*. This is yet another reason for suggesting that the true affinities of *Vaucheria* are to be found with the Xanthophyceae.

REFERENCES

Caulerpa. ARWIDSSON, T. (1930). *Svensk bot. Tidskr.* **24**, 263.
Caulerpa. BÖRGESEN, F. (1907). *K. danske vidensk. Selsk. Skr.* **7**, 340.
Vaucheria. COUCH, J. N. (1932–3). *Bot. Gaz.* **94**, 272.
Caulerpa. DOSTAL, R. (1929). *Planta*, **8**, 84.
Halimeda. HOWE, M. A. (1907). *Bull. Torrey Bot. Club*, **34**, 491.
Neomeris. HOWE, M. A. (1909). *Bull. Torrey Bot. Club*, **36**, 75.
Halicystis. KORNMANN, P. (1938). *Planta*, **28**, 464.
Codium. SCHMIDT, O. C. (1923). *Bibl. bot., Stuttgart*, **91**, 1.
Caulerpa. SCHUSSNIG, B. (1929). *Öst. bot. Z.* **78**, 1.
Caulerpa. SCHUSSNIG, B. (1939). *Bot. Notiser*, p. 75.
Phyllosiphon. TOBLER, F. (1919). *Jb. wiss. Bot.* **58**, 1.
Codium. WILLIAMS, M. (1925). *Proc. Linn. Soc. N.S. Wales*, **50**, 98.

CHAPTER V

CHLOROPHYCEAE (cont.) (CONJUGALES, CHARALES), XANTHOPHYCEAE (HETEROKONTAE), BACILLARIOPHYCEAE, CHRYSOPHYCEAE, CRYPTOPHYCEAE, DINOPHYCEAE

CHLOROPHYCEAE

CONJUGALES

The members of this group are somewhat distinct from the other groups of the Chlorophyceae that have already been described and at one time they were classed in a separate division, the Akontae. As their pigmentation and metabolism are fundamentally the same, however, it would seem desirable to abandon this arrangement. Their peculiar reproduction suggests that they were evolved at a very early stage from one of the simpler orders of the Chlorophyceae. The order is subdivided into two distinct divisions, the Zygnemaceae which are filamentous and the Desmidiaceae most of which are not, although recently some desmids have been classed with the Zygnemaceae.

*ZYGNEMACEAE: Spirogyra (*spiro*, coil; *gyra*, curved). Figs. 67, 68.

The unbranched filaments are normally free-living although attached forms are known, e.g. *S. adnata*, and they form slimy threads which are known as "Water-silk" or "Mermaid's tresses". These grow in stagnant water and are most abundant in either the spring or autumn, the latter phase being due to the germination of a percentage of the spring zygospores. Each cell contains one or more chloroplasts possessing either a smooth or serrate margin and arranged in a characteristic parietal spiral band. The single nucleus is suspended in the middle of the large central vacuole by means of protoplasmic threads that radiate out to the parietal protoplasm. The chloroplasts, which may occasionally be branched, are T- or U-shaped in cross-section and contain numerous pyrenoids which project into the vacuole on the inner side, the majority of the pyrenoids arising *de novo* at cell division. The cell wall is thin and composed, according to some investigators, of two cellulose layers,

whilst others maintain that there is only an inner cellulose layer
with an outer cuticle. The whole filament is enclosed in a mucilage
sheath of pectose. Any cell is capable of division, and vegetative

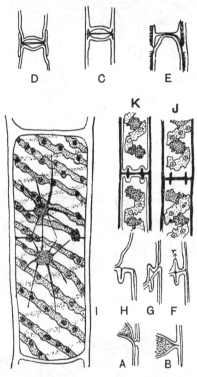

Fig. 67. *Spirogyra.* A, B, cell disjunction (diagrammatic). C–E, cell disjunction in *S. colligata.* F–H, *S. Weberi,* cell disjunction by replicate fragmentation. *r* = replication of septum. I, vegetative structure and cell division,
S. nitida (×266). J, K, cell disjunction and development of replicate septa.
(A–H, J, K, after Fritsch; I, after Scott.)

reproduction by fragmentation is exceedingly common, three
methods having been described:

(*a*) The septum between two cells splits and a mucilaginous
jelly develops in between, so that when one cell subsequently develops a high turgor pressure the cells become forced apart.

(*b*) Ring-like projections develop on both sides of a septum and
the middle lamella dissolves. Then the rings of one cell evaginate

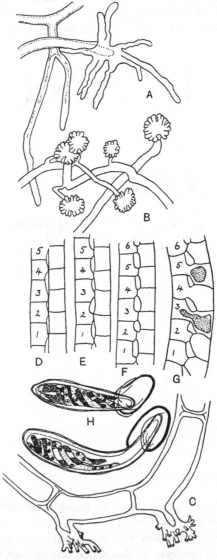

Fig. 68. *Spirogyra*. A, B, rhizoid formation in *S. fluviatilis*. C, rhizoids and haptophores of *S. adnata*. D–G, stages in conjugation, *S. varians*. H, germination of zygospore in *S. neglecta*. (A, B, after Czurda; C, after Delf; D–G, after Saunders; H, after Fritsch.)

and force the cells apart whilst the rings of the other cell evaginate after separation (*replicate fragmentation*) (cf. fig. 67 F–H).

(*c*) The septum develops an ⊥ piece and then when the wall inverts, due to increased turgor, the ⊥ piece is slipped off and the two cells come apart (cf. fig. 67 C–E).

When two filaments touch they may form joints or *genicula-tions*, adhesion being brought about by a mucilaginous secretion produced by the stimulation of the contact. The formation of such geniculations, however, has no connexion with reproduction.

Sexual reproduction is secured by the process of conjugation, the onset of which is brought about by a combination of certain internal physiological factors combined with the pH of the external medium. It commonly takes place during the spring phase and then the threads come together in pairs, but either one or more than two filaments may also be involved. The threads first come together by slow movements, the mechanism of which may be connected with the secretion of mucilage; then they become glued together by their mucilage and later young and recently formed cells in both filaments put out papillae. These papillae meet almost immediately, elongate, and push the threads apart. Normally one of the threads produces male gametes and the other female, but occasionally the filaments may contain mixed cells. The papillae from male cells are usually longer and thinner than those from the female cells and so they can fit inside the latter. The conjugating cells accumulate much starch, the nuclei decrease in size and the wall separating the papillae breaks, thus forming a conjugation tube. The whole process so far described forms the *maturation phase* which is now followed by the *phase of gametic union*. Contractile vacuoles, which make their appearance in the cytoplasmic lining, remove water from the central vacuole and so cause the protoplasm of the male cells to contract from the walls. The male cytoplasmic mass then migrates through the conjugation tube into the female cell where fusion of the two masses takes place and this is then followed by contraction of the female cytoplasm, though in the larger species it may contract before fusion. Fusion of the two nuclei may be delayed for some time, but in any case the male chloroplasts degenerate. The process described above is known as *scalariform* conjugation, and it includes certain abnormal cases where cells produce more than one papilla or where the papillae are crossed. In some monoecious species, however, *lateral* conjugation occurs, the processes being put out from adjoining cells on the same filament.

The last phase to be described is that of *zygotic contraction* which is brought on by further action of the contractile vacuoles, after which a thick three-layered wall develops around the zygote, the middle layer or mesospore frequently being highly sculptured. The zygospore occasionally germinates almost at once, thus producing plants that account for the autumn maximum, but it is usually dormant until the following spring. Meiosis takes place when the zygote germinates and four nuclei are formed of which three abort, the plants thus only exhibiting the haploid generation. A two-celled germling is formed, the lower cell being relatively colourless and rhizoidal in character. Filaments of two different species have been known to fuse, the form of the hybrid zygospore being determined by the characters of the female thread. Azygospores, which have arisen parthenogenetically, and akinetes also form other means of reproduction.

ZYGNEMACEAE: *Zygogonium* (*zygo*, yoked; *gonium*, angle). Fig. 69.

The commonest species of this genus, which is sometimes regarded as a subsection of the genus *Zygnema*, is the terrestrial *Z. ericetorum*. The cells of this species each contain a single axile chloroplast, whilst in *Zygnema*, of course, there is a pair of very characteristic stellate chloroplasts (fig. 69 A). At low temperatures the walls develop a very thick cellulose layer, whilst the sap is coloured violet by phycoporphyrin, especially when the threads are subjected to strong light. Sexual reproduction is rare but when it does occur the gametes are formed from only a part of the protoplasm. In an Indian species azygospores are apparently the only means of reproduction and even these are scarce. Aplanospores and akinetes are commonly formed, and there is one abnormal form growing on Hindhead heath which only exists in the akinete stage.

ZYGNEMACEAE: *Mougeotia* (after J. B. Mougeot, a French botanist). Fig. 70.

The filaments of the different species are commonly unbranched, although they may occasionally possess short laterals. The chloroplast is a flat axile plate lying in the centre of the cell and orientated according to the light intensity, whilst the nucleus is to be found in the centre of the cell on one side of the chloroplast. Fragmentation takes place by method (*a*) as described for *Spirogyra* (cf. above), and knee joints or geniculations are also common. At conjugation the gametes are formed from only part of the cell protoplast as in

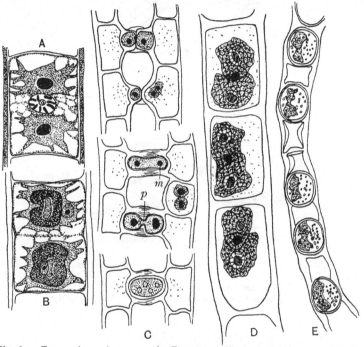

Fig. 69. *Zygogonium ericetorum.* A, *Zygnema stellinum,* cell and nucleus before division (× 500). B, the same, after division (× 500). C, *Zygogonium,* stages in conjugation. *m* = male nucleus, *p* = conjugation process. D, terrestrial form (× 1065). E, aplanospores formed from drying up of filament (× 542). (A, B, after Cholnoky; C–E, after Fritsch.)

Zygogonium, fusion taking place either by way of papillae or through a geniculation. The zygote is cut off by new walls and so becomes surrounded by two or four sterile cells depending on where the zygospore has been formed. Most of the species are isogamous but anisogamy is known in *Mougeotia tenuis.* Reproduction by means of thick-walled akinetes and parthenospores occurs commonly, at least five species having only the latter mode of propagation.

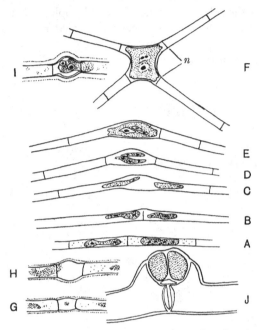

Fig. 70. *Mougeotia.* A–E, *M. mirabilis*, stages in conjugation through loss of cell wall. F, normal conjugation in *M. mirabilis*. *n* = new walls cutting off zygote. G–I, stages in lateral conjugation of *M. oedogonioides*. J, two azygotes in *M. mirabilis*. (A–F, after Czurda; G–I, after Fritsch; J, after Kniep.)

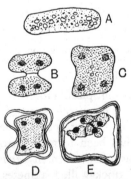

Fig. 71. *Mesotaenium.* A, plant of *M. De Greyi*. B–E, conjugation of *Cylindrocystis Brebissonii* and germination of zygospore. (After Fritsch.)

DESMIDIACEAE: *Mesotaenium* (*meso*, middle; *taenium*, band). Fig. 71.

This is an example of one of the *saccoderm* desmids, which as a group are characterized by a smooth wall in one complete piece and without any pores. The rod-shaped cells of *Mesotaenium* are single, have no median constriction, and are circular in transverse section. The chloroplast is a flat axile plate containing several pyrenoids, whilst in some species the presence of phycoporphyrin imparts a violet colour. The inner cell wall is composed of cellulose and the outer of pectose. Multiplication takes place by cell division, the daughter cells being liberated by dissolution of the middle lamella after a constriction has been formed, though in some cases this may not occur until a number of cells have been enclosed in a common mucilaginous envelope. Sexual reproduction is by means of conjugation, two processes being put out just as in the filamentous forms: these unite and then the middle septum breaks down so that the two protoplasts can meet in the centre, after which the conjugation tube may widen. The thick-walled zygote divides twice, the first division being heterotype, whilst in one species the divisions result in two macro- and two micronuclei. It is from these divisions that either two or four new individuals arise. The species are to be found in upland pools, peat bogs or on the soil.

*DESMIDIACEAE: *Closterium* (enclosed space). Figs. 72, 73.

This genus is an example of one of the *placoderm* desmids, a group that is commonly characterized by the highly perforated cell wall composed of two parts.

The curved cells have attenuated apices with a vacuole in each apex which contains crystals of gypsum that appear to have no physiological function and are probably purely excretory. The pores are arranged in rows in narrow grooves, cell movement being secured by the exudation of mucilage through large pores near the apices. Each semi-cell has one axile chloroplast which is in the form of a curved cone with ridges on it, whilst in transverse section it either has the appearance of a hub with radiating spokes or else looks like a coarsely cogged wheel. Cell division is peculiar and takes place by one of two methods producing either (*a*) connecting bands which appear as striae in the older semi-cells or (*b*) girdle bands (cf. fig. 73). At conjugation, papillae from the two cells meet or else the naked amoeboid gametes fuse immediately outside the cells, whilst in *C. parvulum* there is some evidence of sexual differentiation. After the gametes have fused two of the chloroplasts degenerate and the zygospore on germination divides twice,

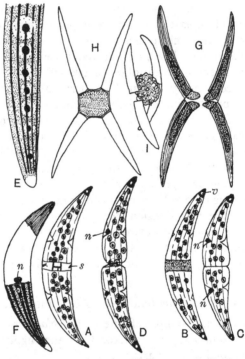

Fig. 72. *Closterium*. A–D, *C. Ehrenbergii*, stages in cell division. n=nucleus, s=septum, v=vacuole. E, *C. lanceolatum*, chloroplast structure. F, *Closterium* sp. showing structure. n=nucleus. G, *C. lineatum*, first stage in conjugation. H, *C. rostratum* var. *brevirostratum*, zygospore formation, second stage. I, *C. calosporum*, mature zygospore. (A–H, after Fritsch; I, after Smith.)

during which meiosis takes place. Two daughter cells are then formed, each containing one chloroplast and two nuclei, but one of the latter subsequently degenerates. The genus is wholly fresh water.

Many of the desmids are planktonic and possess modifications, e.g. spines, which may be regarded as adaptations to this mode of existence. The group is extremely widespread, though it is absent from the Antarctic and is scarce in waters containing much lime, the individual species thriving best in soft or peaty waters. The most favourable seasons for their development are the late spring and

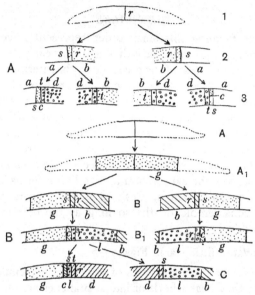

Fig. 73. *Closterium.* Diagrams to explain cell division in species of *Closterium* with (B) and without (A) girdle bands. The different segments of the wall are indicated by shading. 1, 2, 3, and A, B, c = the successive generations. The individuals in 1 and A have each arisen from a zygote and have not undergone division. *a, b, d* = semi-cells of various ages; *c* = the connecting band demarcated by the two sutures *s*, of the previous generation, and *t*, of the present; *g, l* = girdle bands developed before (*g*) and after division; *s* = suture between young and older semi-cells; *r* = the line of the next division. (After Fritsch.)

early summer and their resistance in the vegetative state to adverse conditions would seem to be very great. The evidence suggests that, as a group, they have been evolved from filamentous ancestors, possibly by over-specialization of the process of fragmentation.

REFERENCES

Zygnemaceae. CZURDA, V. (1931). *Beih. bot. Zbl.* 48/2, 238.
Spirogyra, Zygnema. CZURDA, V. (1933). *Beih. bot. Zbl.* 50/1, 196.
Zygogonium. HODGETTS, W. J. (1918). *New Phytol.* 17, 238.
General. LEFÈVRE, M. and MANGUIN, E. (1938). *Rev. gén. Bot.* 50, 501.
Spirogyra. LLOYD, F. E. (1926). *Trans. Roy. Can. Inst.* 15, 151.

108 CHLOROPHYCEAE

Spirogyra. Lloyd, F. E. (1926). *Trans. Roy. Soc. Can.* 3rd series, **20**, 75.
Spirogyra. Lloyd, F. E. (1928). *Protoplasma*, **4**, 45.
General. West, G. S. (1915). *J. Bot.* **53**, 73.
Zygogonium. West, G. S. and Starkey, C. B. (1915). *New Phytol.* **14**, 194.

*CHARALES

The plants forming this small order represent a very highly specialized group that must have diverged very early in the course of evolution from the rest of the green algae, the intermediate forms subsequently being lost. They are characterized in that they lack asexual reproduction and possess very complex sexual reproductive organs. The young plants develop from a protonemal stage, the erect plants having a structure which is more elaborate than any type so far described, whilst the thallus is also frequently lime encrusted. The group is very ancient because fossil members are found from almost the earliest strata. The living forms are widely distributed in quiet waters, fresh or saline, where they may descend to considerable depths so long as the bottom is either sandy or muddy.

*Nitella (*nitella*, a little star). Figs. 74–77.

The plants have the appearance of miniature horsetails (*Equisetum*) because they bear whorls of lateral branches arising from the nodes. The nodes are formed by a transverse layer of cells in contradistinction to the internodes, which consist of one large cell whose individual length may extend up to 25 cm. in *Nitella cernua*. The height of the different species varies up to 1 m., growth being brought about by an apical cell which cuts off successive segments parallel to the base. Each new segment divides transversely into two halves, the upper developing into a node and the lower into an internode (fig. 75 B). Branches, both primary and secondary, are formed by the peripheral cells of the nodes protruding to form new apical cells, but these soon cease to grow after the branch has reached a short length. At the basal node of the main plant branches of unlimited growth are produced: these arise on the inner side of the oldest lateral in the whorl, thus producing a fictitious appearance of axillary branching. Multicellular branched rhizoids with oblique septa function as absorption organs and also serve for anchorage. The rhizoids develop from the lowest node of the main axis, but every node is potentially capable of producing them though normally the presence of the stem apex inhibits their appearance but if this is cut off they will then develop. This behaviour is very suggestive of an auxin control similar to that

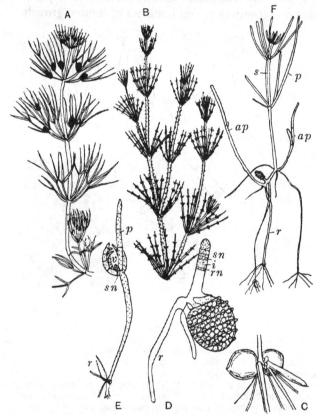

Fig. 74. Charales. A, *Nitella batrachosperma.* B, *Chara hispida.* C, underground bulbil of *C. aspera.* D, germinating oospore. E, protonema of *C. fragilis.* F, young plant of *C. crinita.* *ap*=accessory protonema, *i*=internode, *p*=protonema, *r*=rhizoids, *rn*=rhizoid node, *s*=shoot, *sn*=stem node, *v*=initial of young plant. (After Fritsch.)

found in the higher plants. The cells, which have a cellulose membrane, contain discoid chloroplasts without any pyrenoids together with one nucleus. Cytoplasmic streaming is very readily observed, especially in the internodal cells. Sexual reproduction is by means of a characteristic oogamy where light intensity plays a part in determining the production of the sex organs. The species are either dioecious or monoecious, in which latter case the oogonia and antheridia are juxtaposed, the oogonia being directed upwards and the antheridia downwards, both organs usually appearing on secondary lateral branches of limited growth.

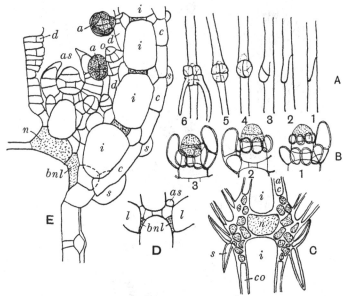

Fig. 75. Charales. A, 1–6, successive stages in development of root node of *Chara aspera*. 1, double foot joint. 2, dilation of toe of upper foot. 3, toe portion cut off. 4, 5, subdivision of toe cell. 6, rhizoids growing out. B, 1–3, successive growth stages of apex of *Nitella*. In 1 apical cell is undivided, in 2 it has divided, in 3 the lower cell has divided into an upper node and a lower internode. C, *C. hispida*, node with stipules. D, *N. gracilis*, longitudinal section of node. E, *C. fragilis*, branch at node with axillary bud. *a* = antheridium, *ac* = ascending corticating cells, *as* = apex of side branch, *bnl* = basal node of branch (*l*), *c* and *co* = cortical cells, *d* = descending cortical cells, *i* = internodal cell, *n* = nodal cell, *o* = oogonium initial, *s* = stipule. (A, B, after Grove; C–E, after Fritsch.)

Antheridia. Fig. 76.

The apical cell of the lateral branch cuts off one or two discoid cells at the base and then becomes spherical. The upper spherical cell divides into octants and this is followed by two periclinal divisions after which the whole enlarges and the eight peripheral cells develop carved plates (*shields*), thus giving the wall a pseudo-cellular appearance. At maturity these peripheral cells acquire brilliant orange contents. The uppermost discoid basal cell protrudes somewhat into the hollow structure formed as described above. The middle segment of each primary diagonal cell now develops into a rod-shaped structure, the *manubrium*, which bears at its distal end one or more small cells, the *capitula*; every one of

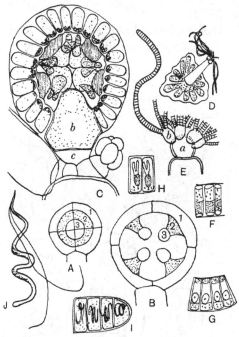

Fig. 76. Charales. A, B, stages in development of antheridium of *Chara*. 1–3, segments and cells to which they give rise. C, section of almost mature antheridium of *Nitella flexilis*. *b* = flask cell, *c* = extra basal cell. D, *C. tomentosa*, single plate with manubrium and spermatogenous threads. E, *C. tomentosa*, apex of manubrium with spermatogenous threads. *a* = primary head cell, *b* = secondary head cell. F–I, *C. foetida*, stages in formation of antherozoids in spermatogenous threads. J, mature antherozoid. (A, B, after Goebel; C–E, J, after Grove; F–I, after Fritsch.)

these produces six secondary capitula from each of which arises a forked spermatogenous thread containing 100–200 cells. These antheridial cells each produce one antherozoid, an elongate body with two flagellae situated just behind the apex. The complete structure has been regarded as one antheridium, whilst another view regards the octants as laterals, the manubrium as an internode, the capitula as a node and the spermatogenous threads as modified laterals, so that on this basis the antheridia are one-celled and conform to the normal structure of the majority of the antheridia in the green algae. This second interpretation, if it is correct, helps considerably in understanding this peculiar group.

Oogonia. Fig. 77.

The apical cell of the lateral branch divides twice giving rise to a row of three cells, the uppermost cell developing into the oogonium

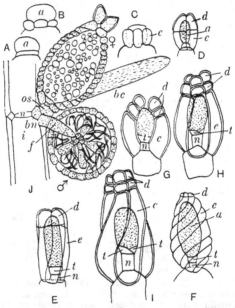

Fig. 77. Charales. A–F, *Chara vulgaris*, stages in formation of oogonium. A, first division. B, C, division of periphery to form envelope cells. D, coronal cells cut off. F, mature oogonium. G–I, *Nitella flexilis*, stages in formation of oogonium. J, fertile branch of *C. fragilis*. *a* = oogonium, *bc* = bract cell, *bn* = branch nodal cell, *d* = coronal cells, *e* = envelope cells, *f* = flask cell, *i* = internode, *n* = nodal cell, *t* = turning cell, *os* = oogonium stalk cell. (After Grove.)

whilst the lowest forms a short stalk. The middle cell cuts off five peripheral cells which grow up in a spiral fashion and invest the oogonium, each one finally cutting off two small *coronal* cells at the apex. The oogonial cell cuts off three cells at its base and it is maintained that these, together with the oogonium, represent four

CHARALES

octants, only one of which develops to maturity. When mature, the investing threads part somewhat to form a neck, and the apex of the oogonium gelatinizes in order to permit the antherozoids to enter. After fertilization the zygote nucleus travels to the apex of the oospore and a coloured cellulose membrane is excreted around it, whilst the oogonium wall, together with inner walls of the investing threads, thicken and silicify. Four nuclei are formed by two successive divisions of the zygote nucleus, meiosis taking place during this process. One of these nuclei becomes cut off by a cell wall whilst the other three degenerate. The small cell so formed then divides and two threads grow out in opposite directions, one a rhizoid, the other a protonema. The cell next to the basal cell of the protonema divides into three cells, the upper and lower forming nodes which become separated by elongation of the middle cell (fig. 74 D–F). The lower node develops rhizoids whilst the upper produces a whorl of laterals from all the peripheral cells except the oldest, which instead forms the apex of the new plant. The mature plant is therefore morphologically a branch of the protonema. Vegetative reproduction can take place from secondary protonemata which develop from the primary rhizoid ring or else from dormant apices.

Chara (of a mountain stream). Figs. 74–77.

This genus is very similar to *Nitella* in its method of reproduction, but the plants are usually larger and coarser as a result of lime encrustation, whilst the stem is corticated, the corticating cells arising from the basal nodes of the short laterals, one thread growing up and another down.

*XANTHOPHYCEAE

As a group the Xanthophyceae exhibit considerably less differentiation than the Chlorophyceae. Two of the most characteristic features are the replacement of starch as a food reserve by oil and a greater quantity of xanthophyll in the plastids, although the actual amount of the latter is partially dependent upon the external conditions. The pigment turns blue-green when the cells are heated in concentrated hydrochloric acid and this forms a convenient test for distinguishing them from the Chlorophyceae. The walls are frequently in two equal or unequal portions which overlap, their composition being principally of a pectic substance although some cellulose may occasionally be present. The motile bodies contain more than one chloroplast and are further characterized by two unequal flagellae, the longer one often possessing delicate cilia. The Xanthophyceae exhibit very little regularity in the formation of reproductive bodies. Sexual reproduction is rare and in the few known examples is always isogamous, the principal mode of reproduction being by means of zoospores and aplanospores. The majority of the species are confined to fresh water. It would seem that they have a motile unicell ancestry, the chief interest of the group being the manner in which evolution has taken place along lines parallel to those found in the Chlorophyceae. As a result there exists a set of analogues which, so far as general morphology is concerned, bear so much resemblance to chlorophycean groups that these forms are classed as Heterochloridales, Heterococcales, Heterosiphonales and Heterotrichales.

HETEROCHLORIDACEAE: *Chloramoeba* (*chlor*, green; *amoeba*, changing). Fig. 78.

This is a naked unicell which is analogous to certain members of the Volvocales, e.g. *Dunaliella*. The cells multiply by longitudinal division, but under adverse conditions ellipsoidal cysts with large oil globules are developed and these form a resting stage.

HETEROCAPSACEAE: *Botryococcus* (*botryo*, cluster; *coccus*, berry). Fig. 79.

This fresh-water genus represents one of the palmelloid analogues of the Chlorophyceae, the principal species, *B. Braunii*, forming an oily scum on ponds and lakes in spring and autumn, whilst in late summer the cells are often coloured red by haematochrome. The colonies vary greatly in shape, the cells being radially arranged into spherical aggregates that are connected in a reticular fashion by tough, hyaline or orange-coloured strands belonging to the lamellated mucous envelope. The individual cells are surrounded by a thin membrane that becomes evident when they are squeezed out of their envelopes as sometimes happens. Each cell is enclosed in a funnel-shaped mucilage cup composed of several layers and prolonged at the base into a thick stalk. In old colonies the mucilage envelope swells up so that the cup structure is obscured, but although the sheath is so predominant nevertheless its origin is not clearly known. The cells multiply by longitudinal division, whilst asexual reproduction by means of zoospores has also been recorded though it requires confirmation. Normally reproduction is secured by means of aplanospores, of which two to four are

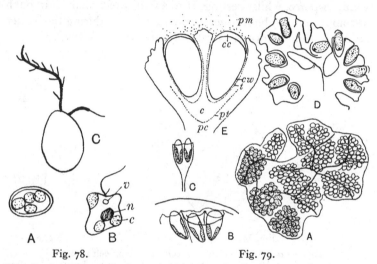

Fig. 78.   Fig. 79.

Fig. 78. A, *Chloramoeba heteromorpha*, cyst. B, the same, motile phase. *c* = chloroplast, *n* = nucleus, *v* = vacuole. C, flagellum structure in *Monocilia*. (After Fritsch.)

Fig. 79. *Botryococcus Braunii*. A, colony (× 300). B, portion of colony showing cells in their mucilage envelope. C, two cells enclosed in the parent cup. D, portion of colony enlarged (× 780). E, two cells arranged diagrammatically to show structure. *c* = cup, *cc* = cell cap, *cw* = cell wall, *pc* = parent cell, *pm* = pectic mucilage, *pt* = parent thimble, *t* = thimble. (A, after Smith; B, C, after Fritsch; D, E, after Blackburn.)

produced in each cell. The colonies decay very slowly, and one of the principal interests of the genus is the recent discovery that boghead coal is composed very largely of this organism, whilst the fossil genera *Pila* and *Reinschia* hardly differ from the living *Botryococcus Braunii*.

HALOSPHAERACEAE: *Halosphaera* (*halo*, salt; *sphaera*, sphere). Fig. 80.

The large, free-floating spherical cells possess one nucleus which is suspended either in the central vacuole or else in the parietal cytoplasm where it is associated with numerous discoid chloroplasts. A new membrane is formed internally and then the old one ruptures, but as the latter may still persist outside one can often see what appears to be a multi-layered sheath. Reproduction can take place by means of zoosporic swarmers but these may be replaced by aplanospores, whilst resting cysts are also recorded. Although most abundant in the warmer oceans, especially during the winter months, its life history is as yet only imperfectly known.

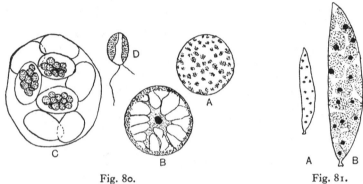

Fig. 80.   Fig. 81.

Fig. 80. *Halosphaera viridis*. A, mature cell. B, young cell in optical section. C, mature cell with aplanospores. D, swarmer. (A–C, after Fritsch; D, after Dangeard.)

Fig. 81. *Characiopsis saccata*. A, plant. B, probable swarmer formation. (After Fritsch.)

CHLOROTHECIACEAE: *Characiopsis* (like *Characium*). Fig. 81.

The very name of this genus indicates that it is an analogue to the genus of similar name in the Chlorophyceae. The plants, which are epiphytic, solitary or gregarious, vary much in shape, even in pure culture, and they develop from a short stalk with a basal mucilaginous cushion. The wall, composed of cellulose and pectins, is in two unequal portions, the smaller upper part forming a lid which is detached at swarmer formation whilst in one species the lower part bears internal processes. Although the young cells are uninucleate and contain one or more chloroplasts the adult cells are multinucleate containing eight to sixty-four nuclei. Reproduction is either by means of zoospores (eight to sixty-four per cell) or else by means of thick-walled aplanospores, which in one species are said to give rise to motile gametes, although this is a feature that requires further investigation.

*TRIBONEMACEAE: *Tribonema* (*tribo*, thin; *nema*, thread). Fig. 82.

This is a filamentous analogue to a form such as *Microspora* (cf. p. 46) with which it is frequently confused. *T. bombycina* sometimes appears in sheets covering ponds and pools and if these dry up they form an algal "paper". The unbranched threads are composed of cells possessing walls of two equal overlapping halves, with the result that the filaments are open-ended and tend to dissociate into H pieces. At cell division a new H piece arises in the

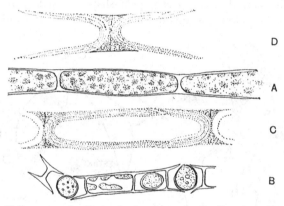

Fig. 82. *Tribonema*. A, *T. bombycina* (×450). B, *T. minus*, hypnospores. C, D, construction of H piece in *T. bombycina* as shown after treatment with KOH (×675). (A, C, D, after Smith; B, after Fritsch.)

centre and the two halves of the parent cell separate, somewhat as in the Desmidiaceae. Each cell contains one nucleus, although *Tribonema bombycina* may have two together with two or more parietal chloroplasts. Asexual reproduction is by means of zoospores (two to four per cell) which are liberated by separation of the two halves of the cell. On coming to rest the zoospore elongates and puts out an attachment process, and in this state it much resembles *Characiopsis*. Aplanospores (one to two per cell) and akinetes, which are formed in chains, also act as additional means of propagation, whilst sexual reproduction has been seen only once when some of the motile bodies came to rest first and were surrounded by other motile gametes. Iron bacteria sometimes live symbiotically with this alga and colour it yellow or brown from ferric carbonate. This substance controls the *p*H of the water and thus acts as a local buffer for the alga whilst the bacteria obtain their oxygen requirements from the *Tribonema*.

BOTRYDIACEAE: *Botrydium* (a small cluster). Fig. 83.

This genus belongs to the Heterosiphonales and is analogous to a form such as *Protosiphon*, the commonest species, *Botrydium granulatum*, being frequently confused with it, especially as these two plants are often associated on areas of drying mud. *B. granulatum* makes its appearance during the warmer part of the year when it is seen that the green, pear-shaped vesicles are rooted by means of colourless dichotomously branched rhizoids. The

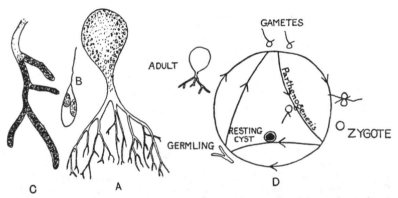

Fig. 83. *Botrydium granulatum*. A, plant. B, swarmer. C, cyst formation. D, diagram of life cycle. (A–C, after Fritsch; D, after Miller.)

membrane is composed of cellulose and the lining cytoplasm contains numerous nuclei scattered throughout it, whilst the chloroplasts, containing pyrenoid-like bodies, are confined to the aerial part. The shape of the vesicle is influenced by the environment, the shade forms being elongate or club-shaped. In *B. Wallrothii* the unbranched vesicle is covered with lime whilst in *B. divisum* it is branched but without lime. When the plants are submerged reproduction takes place by means of numerous zoospores which are set free by gelatinization of the vesicle apex, but when the plants are only wet but not submerged aplanospores are formed instead. Under dry conditions each vesicle develops into a single cyst (macrocyst) or into several multinucleate spores (sporocysts), or else the contents migrate to the rhizoids and there form several cysts (rhizocysts) which, when conditions are again favourable, either germinate directly to a new plant or else give rise to zoospores. In *B. granulatum* it is estimated that about 40,000 isogametes are formed in each vesicle, but as the plant is monoecious many fuse either in pairs or threes, rarely fours, before they are liberated. Those that do not fuse develop parthenogenetically, although the stage at which meiosis occurs is not yet known. The life cycle can be tentatively represented as in fig. 83 D.

*BACILLARIOPHYCEAE (DIATOMACEAE)

Figs 84, 85

These unicellular algae are abundant as isolated or colonial forms in marine or fresh-water plankton and also as epiphytes on other algae and plants. They form a large proportion of the bottom flora of lakes and ponds and occur widely on salt marshes, although certain diatoms are said to be very sensitive to the degree of salinity in the medium. In the colonial forms the cells are attached to each other by mucilage or else they are enclosed in a common mucilaginous envelope. The plants have characteristic silicified cell walls which are built up on a pectin foundation and are highly sculptured. Each shell (*frustule*) is composed of two halves varying much in shape, the older (*epitheca*) fitting closely over the younger (*hypotheca*), each half being composed of a valve together with a connecting band, the latter forming the overlapping portion. The Diatomaceae are divided into two groups, the Pennatae and Centricae, the former having intercalary bands as well as the connecting bands. A simple way of distinguishing between these two groups is that the Pennatae have the shape of date boxes and the Centricae that of pill boxes. The marks or striae on the frustules are composed of rows of dots which represent small cavities, and these are so fine that they are employed in testing the resolving power of microscopes. The Pennatae have the striae arranged in series with either a plain area in between (*pseudoraphe*) or else a slit that varies in form and structure (*raphe*). In the Centricae these structures are absent and the striae are arranged radially. The raphe is connected with movement, as only those forms possessing one have the power of locomotion, and although the mechanism is not completely understood it would seem to be connected with friction caused by

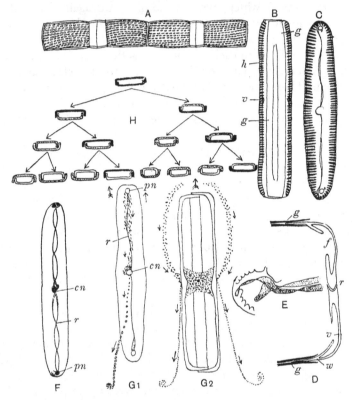

Fig. 84. Bacillariophyceae. A, *Melosira granulata* (Centricae) (× 624). B, *Pinnularia viridis* (Pennatae), girdle view. C, same, valve view. D, *P. viridis*, union of valve and parts of adjacent girdle bands. E, *P. viridis*, termination of the two parts of the raphe in the polar nodule. F, *P. viridis*, diagrammatic view showing the two raphes. G, movement of *P. viridis* as shown by sepia particles. 1, in valve view; 2, in girdle view. H, diagram to illustrate successive diminution in size of plant. The half-walls of the different generations are shaded appropriately. *cn* = central nodule, *f* = foramen, *g* = girdle, *h* = hypotheca, *pn* = polar nodule, *r* = raphe, *v* = valve, *w* = wall of valve. (A, H, after Smith; B–G, after Fritsch.)

the streaming of protoplasm. Streams of mucilage pass from the anterior polar nodule down to the centre of the plant body where it masses and then spreads out posteriorly in the form of a fine thread (fig. 84 G). Each cell is surrounded by a cytoplasmic lining with a bridge between the two halves of the shell in which the nucleus is commonly to be found. The chloroplasts are parietal, olive green to brown, the principal colouring matter being isofucoxanthin, whilst pyrenoids may be present or absent. The product of photosynthesis is a fatty oil. The pelagic forms frequently possess outgrowths which must be regarded as adaptations to their mode of existence. Cell division normally occurs at night time, and when the nucleus and protoplast have divided new valves are formed

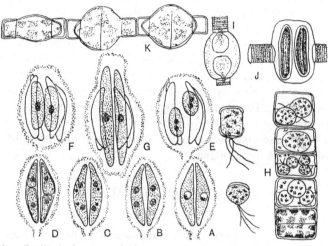

Fig. 85. Bacillariophyceae. A–G, auxospore formation by two cells in the pennate diatom, *Cymbella lanceolata*. A, synaptic contraction. B, after first mitosis. C, second mitosis with functional and degenerating pairs of nuclei. D, division of each protoplast into two uninucleate gametes. E, young zygotes. G, zygotes elongated to form auxospores. H, microspore formation in *Melosira varians* (× 600). I, J, auxospore formation in *Rhabdonema arcuatum*. K, asexual auxospores in *M. varians*. (A–H, K, after Smith; I, J, after Fritsch.)

inside and then the parent connecting bands separate. One individual thus becomes smaller and smaller because the size of the new valve is fixed by the silica contained in the wall of the old valve and in five months there may be a decrease of three-fifths to two-thirds of the length until finally the shrinkage is compensated for by auxospore formation (fig. 84 H). However, a long time elapses before this rejuvenation is necessary and so auxospore formation is relatively rare.

At auxospore formation in the Centricae the two halves of the shell are thrust apart by enlargement of the protoplast, which becomes enveloped in a slightly silicified pectic membrane, the *perizonium*. No nuclear division takes place, but fresh valves and connecting bands are formed inside this membrane so that a new and larger individual results. In the Pennatae a union takes place between two naked amoeboid protoplasts that have arisen from two distinct individuals which come together in a common mucilaginous envelope. These are the gametes, and as meiosis occurs during their formation the normal diatom cell must be regarded as diploid (fig. 85). The zygote remains dormant for a time and then elongates in order to form auxospores, the perizonium either being the remains of the zygotic membrane or else formed *de novo*. Isogamy is the normal condition but a few cases of physiological anisogamy are known and also apogamy. In addition to auxospores the Centricae also produce microspores, small rounded bodies with flagellae, but whether these are true gametes has yet to be established because their fate has not been fully studied. Some diatoms are also known to produce resting spores but very little is recorded about these bodies.

CHRYSOPHYCEAE

Fig. 86

This assemblage is principally composed of uninucleate flagellate forms although certain members do exhibit some algal characteristics. Like the Xanthophyceae there is considerable morphological parallelism with the Chlorophyceae indicating that evolution has taken place along the same lines. Sexual reproduction is rare and when it does occur is isogamous, the plants probably all being haploid. They occur most commonly in both fresh or salt water during cold weather. The colour is golden yellow or brown due to the presence of the pigment phycochrysin which is contained in a small number of parietal chromatophores that may also contain pyrenoid-like bodies, although starch as a product of photosynthesis is replaced by oil or leucosin. The motile cells are uni- or biflagellate, and in the latter event one flagellum is beset with fine cilia; one of the flagellae is said to provide forward movement and the other rotation. When an individual has entered the amoeboid state cysts may be produced endogenously and these have silicified walls composed of two equal or unequal parts. The group possesses the following morphological categories:

(*a*) Unicellular motile types, e.g. *Chromulina*.

(*b*) Encapsuled types, either free or epiphytic, e.g. *Dinobryon* spp.

(*c*) Colonial types, e.g. *Synura*.

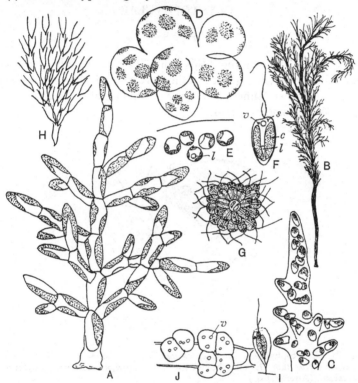

Fig. 86. Chrysophyceae. A, *Phaeothamnion confervicolum*. B, *Hydrurus foetidus*. C, *H. foetidus*, apex showing branching. D, *Phaeocystis pouchetii*. E, the same, portion of plant. *l* = leucosin. F, *Ochromonas mutabilis*. *c* = chloroplast, *l* = leucosin, *s* = stigma, *v* = vacuole. G, *Synura ulvella*. H, *Dinobryon sertularia*, colony. I, *D. marchicum*. J, *Epichrysis paludosa* on *Tribonema*. *v* = vacuole. (After Fritsch.)

(*d*) Dendroid colonies, e.g. *Dinobryon* spp.

(*e*) Rhizopodial or amoeboid types, e.g. *Rhizochrysis*.

(*f*) Palmelloid types, e.g. *Phaeocystis* and *Hydrurus*, the latter being a highly differentiated branched type.

(*g*) Simple filamentous types, e.g. *Phaeothamnion*.

It is suggested that the group is still actively evolving, and that some of the brown types with algal characters may have a relationship with the simpler Phaeophyceae.

CRYPTOPHYCEAE

Fig. 87

Very little is known about this group. They are mostly specialized flagellates with two flagellae but there are a few algal forms, although none of them is filamentous. The morphological types are:

(*a*) *Naked motile unicells.*

(*b*) *Colourless unicells.*

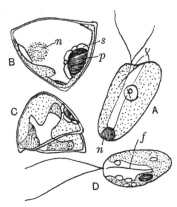

Fig. 87. Cryptophyceae. A, *Cryptomonas anomala*, side view. B–D, *Tetragonidium verrucatum*, D, being the swarmer. *f* = furrow, *n* = nucleus, *p* = pyrenoid, *s* = starch. (After Fritsch.)

(*c*) *Symbiotic unicells* with cellulose walls, e.g. some of the *Zooxanthellae* which are found associated with Coelenterata and Porifera.

(*d*) *Palmelloid* type, e.g. *Phaeococcus*, which is found on salt marsh muds in England.

(*e*) A single *coccoid* type, *Tetragonidium*.

The number of chloroplasts varies, pyrenoids are present, and there is one nucleus in each cell. Reproduction is by means of longitudinal fission but some species also form thick-walled cysts.

DINOPHYCEAE

Fig. 88

This group is predominantly planktonic, naked forms being most abundant in the sea, whilst in fresh waters one commonly finds armoured forms which often have spiny processes that can be regarded as adaptations to their pelagic existence. The majority of the species are motile and characteristically possess two flagellae, one directed forward and one transversely, both commonly lying in grooves and emerging through pores. In one or two cases, however,

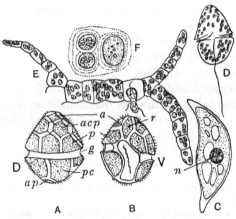

Fig. 88. Dinophyceae. A, *Peredinium anglicum*, dorsal view. B, *P. anglicum*, ventral view. C, *Cystodinium lunare*. D, *Gymnodinium aeruginosum*. E, *Dinoclonium Conradi*. F, *Gloeodinium montanum*. *a* = apical plate, *acp* = accessory plates, *ap* = antapical plates, *g* = girdle, *n* = nucleus, *p* = precingular plates, *pc* = postcingular plates, *r* = rhomboidal plate. (After Fritsch.)

the flagellae may be situated anteriorly. In some forms *ocelli*, which are composed of a spherical lens and a pigment, can be observed; these are presumably connected with the perception of light and they must be regarded as an elaborate development of the ordinary red eye-spot. Two genera also possess nematocysts comparable to those found in hydroids. The numerous disk-like chromatophores are dark yellow or brown in colour and sometimes contain pyrenoids. There is one nucleus and the food reserve is starch and fat, whilst the marine Dinoflagellates are noted for possessing large vacuoles. Multiplication is by means of cell division which takes place either during the motile phase or else during a resting phase. Spherical swarmers of the naked unicell type are also known together with cysts and autospores. The following represent the different morphological types that have been evolved in the course of evolution:

(a) Motile unicells which are either naked or else enclosed in a delicate membrane, e.g. Desmokontae and the unarmoured Dino-flagellates.

(b) Motile unicells with a conspicuous cellulose envelope of sculptured plates and with the flagellae furrows well marked, e.g. armoured Dinoflagellates—*Peredinium, Ceratium.*

(c) Parasitic marine forms which are either ecto- or endo-parasites.

(d) One palmelloid genus, *Gloeodinium.*

(e) Colourless and rhizopodial forms.

(f) Coccoid forms, e.g. Dinococcales.

(g) Filamentous forms, e.g. *Dinothrix, Dinoclonium.*

Recent work has tended to show that there is no real evidence for believing that this group is closely related to the Diatomaceae as was formerly supposed.

REFERENCES

Botryococcus. BLACKBURN, K. and TEMPERLEY, B. N. (1936). *Trans. Roy. Soc. Edinb.* **58**, 841.

Halosphaera. DANGEARD, P. (1932–3). *Botaniste,* **24**, 261.

Diatoms. GEITLER, L. (1930). *Arch. Bot.* **3**, 105.

Diatoms. GEITLER, L. (1932). *Arch. Protistenk.* **78**, 1.

Charales. GOEBEL, K. (1930). *Flora,* **124**, 491.

Diatoms. GROSS, F. (1938). *Philos. Trans.* B, **228**, 1.

Charales. GROVES, J. and BULLOCK-WEBSTER, G. R. (1920–4). *The British Charophyta,* **1**, **2**. Ray Society.

General. KOLBE, R. W. (1927). *Pflanzenforschung,* **7**.

Botrydium. MILLER, V. (1927). *Ber. dtsch. bot. Ges.* **45**, 151.

Dinophyceae. PASCHER, A. (1927). *Arch. Protistenk.* **58**, 1.

Xanthophyceae. POULTON, E. M. (1926). *New Phytol.* **25**, 309.

Xanthophyceae. POULTON, E. M. (1930). *New Phytol.* **29**, 1.

CHAPTER VI

PHAEOPHYCEAE

ISOGENERATAE AND HETEROGENERATAE (EXCLUDING DICTYOTALES, LAMINARIALES AND FUCALES)

*GENERAL

The algae composing this group range from minute disks to 100 m. or more in length and are characterized by the presence of a brown pigment, fucoxanthin; the function of this substance is still not clearly understood but it is probably connected with the absorption of light, though not with its utilization (cf. p. 293). In the older classification the group was customarily divided into three orders: the Phaeosporeae (including the Laminariales) with motile gametes, Cyclosporeae with non-motile asexual bodies and non-motile ova and the Acinetosporeae or Tilopteridales with non-motile asexual bodies, the sexual reproductive organs being either absent or imperfect. In 1917 Kylin suggested the following classification: Tilopteridales (Acinetosporeae), Dictyotales, Laminariales, Fucales and Phaeosporeae, this last group really being a polyglot assembly of distantly related forms. Later Taylor (1922) transferred the Laminariales to the Cyclosporeae and placed the Acinetosporeae in the Phaeosporeae thus leaving only two orders. In 1933 Kylin suggested yet another rearrangement with only three groups based upon the type of alternation of generations, but it is doubtful whether this new classification has any more real significance in so far as phylogenetic relationships are concerned.

(*a*) *Isogeneratae.* Plants with two morphologically similar but cytologically different generations in the life cycle, e.g. Ectocarpaceae, Sphacelariaceae, Dictyotaceae, Tilopteridaceae, Cutleriaceae.

(*b*) *Heterogeneratae.* Plants with two morphologically and cytologically dissimilar generations in the life cycle:

    I. Haplostichineae. Plants with branched threads, which are often interwoven, and without intercalary growth, e.g. Chordariaceae, Mesogloiaceae, Elachistaceae, Spermatochnaceae, Sporochnaceae, Desmarestiaceae.

II. Polystichineae. Plants built up by intercalary growth into a parenchymatous thallus, e.g. Punctariaceae, Dictyosiphonaceae, Laminariales.

(c) *Cyclosporeae*. Plants possessing a diploid generation only, e.g. Fucales. In view, however, of the most recent interpretation of the life history of the Fucales (p. 189) the Cyclosporeae should now be classed with the Heterogeneratae, division Polystichineae.

It would seem impossible to construct a classification of the Phaeophyceae on a satisfactory phylogenetic basis because they would appear to have diverged and converged greatly during the course of evolution. As a group they are very widespread and are confined almost entirely to salt water although *Pylaiella* is sometimes found in brackish water and *Lithoderma* in fresh water. Some of the species commonly exhibit morphological variations and it has been shown that these may depend on (a) season of the year, and (b) nature of the locality. Church (1920) has given us an elaborate account of the morphology of the Phaeophyceae, and he suggested that if a brown flagellate came to rest it could develop in one of three directions to give:

(a) Uniseriate filaments which occupy a minimum area and obtain maximum light energy per unit of area, growth being either distal or intercalary.

(b) A mono- or polystromatic thallus which occupies a maximum area and obtains a minimum light energy per unit of area.

(c) Mass aggregation.

A morphological examination of the brown algae will show that development has taken place along each of these directions, often resulting in plant bodies of a complex construction, and the following types can be recognized among the various species:

(a) Simple filaments (e.g. *Acinetospora*).

(b) Branched filaments (e.g. *Pylaiella*).

(c) Erect filaments arising from a basal portion (e.g. *Myrionema*).

(d) Interwoven central filaments (*cable* type, e.g. *Mesogloia*).

(e) Basal portion only (*reduced* filamentous or cable type, e.g. *Phaeostroma*).

(f) Filaments uniting to form a sphere (*hollow parenchymatous* or modified cable type, e.g. *Leathesia*).

(g) Multiseptation of primary cable type (e.g. *Chorda*).

(*h*) Erect filaments with cortication (*corticated* type, e.g. *Sphacelaria*).

(*i*) Simple or laminate *parenchymatous* thallus (e.g. *Punctaria*).

(*j*) *Improved parenchymatous* structure with internal differentiation of the tissues (e.g. Laminariales).

(*a*) to (*c*) are generally of an ectocarpoid type (like *Ectocarpus*; cf. p. 132) with a single central filament, or else of a mechanically produced cable type when the central filaments are twisted together by wave action to form a rope or cable. Many of these, whether reduced or not, exhibit the condition of heterotrichy similar to that found in the Chaetophorales, but this is a feature that will be discussed elsewhere (cf. p. 263). The thalli may also reach a relatively large size and under these circumstances additional support is obtained as follows:

(1) Increase in wall thickness (*Stypocaulon*) or the production of a firmer cellulose material (*Sphacelaria*).

(2) Twisting and rolling of the threads together.

(3) Development of root branches or *haptera*.

(4) The appearance of descending and ascending corticating filaments.

(5) Multiseptation takes place in a longitudinal direction.

Branching may proceed from any cell, and it frequently takes the form of a regular or irregular dichotomy, although sometimes a spiral phyllotaxis may be found.

The cells vary greatly in size but they always have distinct walls, which are usually composed of cellulose, and although they are uninucleate occasionally they become multinucleate. Plastids are also present, but the green colour is masked by the brown pigment fucoxanthin. This, however, can be removed by boiling and the thallus then takes on a green colour from the chlorophyll, the composition of which is not quite the same as that of the higher plants because chlorophyll *b* is absent (cf. p. 290) whilst xanthophyll is also missing in the higher members, e.g. Fucales. The products of assimilation are alcohols, carbohydrates and oils but no true starch is formed. Hyaline hairs occur in many forms and their function has been variously ascribed as

(1) shock absorbers,

(2) respiratory and absorptive organs,

(3) protection against intense illumination,
(4) protection against epiphytes,
(5) protection against covering by sand or silt,
(6) mucilage organs.

None of the evidence for any of these suggestions is entirely satisfactory, and the whole problem demands further investigation.

Vegetative reproduction may take place by splitting of the thallus or else by the development of special propagules (*Sphacelaria*). Asexual reproduction is commonly secured by means of uni- or biflagellate zoospores which are normally produced in specialized cells or sporangia. In one group (Dictyotales) tetraspores replace the zoospores, these bodies being produced in groups of four in each sporangium on plants which do not bear sexual organs. In yet another group (Tilopteridales) asexual reproduction is by means of uni- to quadrinucleate monospores. The homologies of these monospores have been subject to much speculation and they have been variously regarded as equivalent to

(*a*) propagules of *Sphacelaria*,
(*b*) simple forerunners of tetraspores,
(*c*) degenerate tetraspores,
(*d*) parthenogenetic ova.

The second suggestion is perhaps the most satisfactory in our present state of knowledge, especially when considered in relation to the vegetative characters. Sexual reproduction ranges from isogamy, with both gametes motile and *characteristically* bearing two flagellae *inserted laterally*, through a series in which differentiation first to anisogamy and finally to oogamy can be traced. Only one species (cf. p. 184) is known in which the ova are retained on the parent plant, so that, apart from this exception, fertilization always takes place in the water. The change from isogamy to anisogamy is also accompanied by a corresponding differentiation of the gametangia.

Both unilocular and plurilocular sporangia are commonly found, but the fate of their products varies considerably (cf. p. 247). Most species show an alternation of generations, but this is by no means regular as there may be considerable modifications. Indeed, the alternation in the Ectocarpales is so irregular that it has been

GENERAL

proposed to term the phenomenon a race cycle rather than an alternation of generations, and this would appear to be the better terminology. Furthermore, the two generations are often not the same in size, and commencing from species with equal morphological generations one may have those in which either the sporophyte or gametophyte is dominant down to plants where only the gametophyte or sporophyte is known. A progression in anatomical development can be traced, but it seems almost impossible to do the same when the life histories or reproductive organs are considered. Three principal types of life cycle have, however, been recognized by Kylin:

(*a*) *Fucus* type. Only the diploid sporophytic plant is known, with meiosis taking place at gametogenesis, e.g. Fucales. This type, however, is more apparent than real (cf. p. 189).

(*b*) *Dictyota* type. Meiosis is delayed and two similar morphological generations exist, e.g. Dictyotales, *Nemoderma*, *Lithoderma*, *Ectocarpus siliculosus* (certain areas only), *Pylaiella littoralis*. A modification of this type is found in *Cutleria* where the two generations are of equal significance but morphologically dissimilar, the gametophyte being the larger.

(*c*) *Laminaria* type. Meiosis is delayed but the two generations are wholly *dissimilar*, the sporophyte being dominant whilst the gametophyte is much reduced.

Quite a number of species must now be included in this last category, although the regular alternation may be masked by complications produced by such phenomena as parthenogenetic development of the ova. Those members of the Heterogeneratae (excluding the Laminariales) which exhibit this type of alternation have a fully developed diploid or *delophycée* form which is common in summer, and a much reduced haploid or diploid *adelophycée* stage which usually appears during the winter months in one of the following forms:

(*a*) In a *protonemal* stage which reproduces the large form by means of buds.

(*b*) In a gametophytic *prothallial* stage which reproduces the large form by means of gametes from plurilocular sporangia.

(*c*) In a *plethysmothallial* stage which reproduces the large form by means of swarmers from either unilocular or plurilocular

131

9-2

sporangia. Until recently these were regarded as arrested sporophytes in a juvenile condition.

Fritsch, however, has suggested (1939) that some of these plethysmothalli are really potential gametophytes (prothalli), especially
those dwarf plants which perpetuate themselves by means of
plurilocular sporangia. The term "plethysmothallus" should be
reserved for plants that are diploid and which have arisen from
diploid swarmers produced in plurilocular sporangia on the
macroscopic plants. It has been suggested that this type of alternation should be termed an alternation of vegetation growths rather
than an alternation of generations, but it is also equally satisfactory
to regard it as *heteromorphic alternation.*

REFERENCES

CHURCH, A. H. (1920). Somatic Organisation of the Phaeophyceae. *Oxf.*
    Bot. Mem. no. 10.
FRITSCH, F. E. (1939). *Bot. Notiser*, p. 125.
KYLIN, H. (1933). *Lunds Univ. Årsskr.* N.F., Avd. 2, **29**, no. 7.
TAYLOR, W. R. (1922). *Bot. Gaz.* **74**, 431.
WILLIAMS, J. LLOYD (1925). *Rep. Brit. Ass.* Pres. Address, Sect. K, p. 182.

ECTOCARPALES
(ISOGENERATAE AND HETEROGENERATAE)

*ECTOCARPACEAE: *Ectocarpus* (*ecto*, external; *carpus*, fruit). Figs.
    89, 90.

The plants are composed of uniseriate filaments which are
sparsely or profusely branched. The erect portion is sometimes
decumbent and arises from a rhizoidal base, which in some of the
epiphytic species occasionally penetrates the host, and it is also
possible that there may be one or two examples of mild parasitism.
E. fasciculatus even grows on the fins of certain fish in Sweden, but
the nature of the relationship in this case is not clear. The branches
of some species terminate in a colourless mucilage hair: in young
plants of *E. siliculosus* these hairs are quite long, but later, with
increasing age, they become much shorter through truncation. The
erect filaments have an intercalary growing region, but the
rhizoids increase in length by means of apical growth. Each cell,
which contains one nucleus together with brown disk or band-
shaped chromatophores, possesses a wall that is composed of three
pectic-cellulose layers.

Generally two kinds of reproductive structures are present, the plurilocular and unilocular sporangia, but some species possess a third type, the meiosporangia. The unilocular sporangia always occur on diploid plants and they give rise, after meiosis, to numerous haploid zooids which may either function as gametes or else develop without undergoing a fusion. The sporangia are sessile or stalked and vary in shape from globose to ellipsoid, the mature ones dehiscing through the swelling up of the centre layer in the wall. The plurilocular sporangia, which are either sessile or stalked, range from ovate to siliquose in shape and are to be found on haploid or diploid thalli. In *E. siliculosus* they represent modified

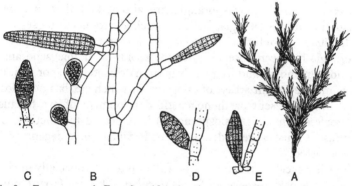

Fig. 89. *Ectocarpus*. A, *E. confervoides*, plant (× 0·44). B, *E. tomentosus*, unilocular and plurilocular sporangia (× 100). C, meiosporangium, *E. virescens*. D, megasporangium, *E. virescens*. E, microsporangium, *E. virescens*. (A, B, original; C–E, after Kniep.)

lateral branches and arise as side papillae from a vegetative cell in the filament. The plurilocular sporangia are divided up into a number of small cells, each one of which gives rise to a zooid and, when ripe, dehiscence takes place by means of an apical pore, the contents either germinating directly or else behaving as gametes. The gametes are usually alike in size but the sex function becomes weaker with age so that relative sexuality is induced, the older and weaker gamete behaving as the opposite sex towards the younger and stronger gamete.

In one species, *E. secundus* (*Giffordia secundus*), there is well-marked anisogamy because there are two types of plurilocular sporangia with large or small loculi that produce zooids which differ

in size, the smallest being the antheridia and the largest, or megasporangia, the oogonia. The contents of the larger megasporangia
are sometimes capable of parthenogenetic development, when they
must be regarded as incipient or degenerate ova. In *E. Padinae*
the unilocular sporangia are absent and there are three kinds of
plurilocular sporangia. One type, which has very small loculi,
represents the antheridia, whilst there are also medium-sized or
meiosporangia, and large or megasporangia. The latter probably
represent the female reproductive organs, but there is, at present,
no definite proof for this hypothesis. It has been suggested that the
meiosporangia may be haploid and the megasporangia diploid in
character, but no cytological data appear to be available. In
E. virescens unilocular sporangia are absent and there are only
meio- and megasporangia, both of which always occur on separate
individuals. No fusion between zooids from the two types of
sporangia has been observed, but the zooids of the megasporangia
are not very mobile and frequently germinate inside the sporangium.
This may represent a case of apogamy in which sex has been lost,
or it may represent parthenogenetic development of ova because
the male organs (the meiosporangia) have ceased to function. In
any case it must be regarded as a type in which some degeneration
has occurred.

The life cycles of the species are full of interest, especially in view
of what has been discovered for *E. siliculosus*. Knight (1929) found
that the plants in the Isle of Man occurred in early spring and late
autumn and were all diploid, the haploid generation being unknown.
They bore unilocular and plurilocular sporangia, the former
producing gametes after a reduction division whilst the latter gave
rise to zoospores. In the Bay of Naples, on the other hand, the
large plants were all haploid and only bore plurilocular sporangia.
The zooids from these behaved as gametes, and after fusion meiosis
commonly took place when the zygote commenced to germinate
because it normally developed directly into a new haploid plant.
Berthold recorded a microscopic form which has since been regarded as diploid because unilocular sporangia were found on it,
but Knight was unable to find any such dwarf plants.

A schema illustrating these features is seen in fig. 90. It has been
suggested that the differences between the plants from the two
localities are due to differences in the tides, light conditions or

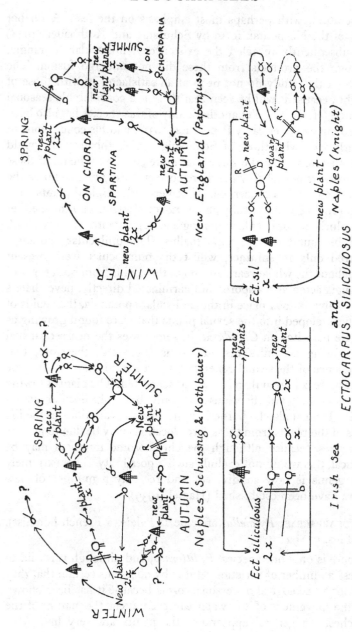

Fig. 90. Diagram to show different life cycles recorded for *Ectocarpus siliculosus* from different localities.

temperature, with perhaps most emphasis on the last. A further study of the Neapolitan form by Schussnig and Kothbauer (1934) has subsequently revealed the existence of unilocular sporangia, although the products from these did not undergo fusion. The results of this study do not fit in at all satisfactorily with those of Knight because it will be seen that there is a considerable seasonal variation (cf. fig. 90). Yet another study of this species has also been carried out in America by Papenfuss (1935), and his conclusions fit in fairly well with those of Schussnig and Kothbauer: It would seem, therefore, that the somewhat more complex schema of these later workers is probably the more correct, at any rate so far as the Neapolitan form is concerned. In America the diploid plants were found growing epiphytically on *Chorda* or *Spartina* and these either bore pluri- or unilocular sporangia independently, or else both could be found on the same thallus. The unilocular sporangia occurred only in summer, whilst the plurilocular were present throughout the whole year. Although the zooids from both types of sporangia acted as zoospores and germinated directly, nevertheless meiosis always took place in the unilocular sporangia, the zooids of which developed into the sexual plants that were found growing as obligate parasites on *Chordaria*, in some cases the nearest asexual plants being 20 miles distant. It is suggested, therefore, that dependence of the sexual generation upon a particular host may be rather more common than is perhaps suspected. The plants growing on *Chordaria* were dioecious and only bore plurilocular game-tangia. It must also be borne in mind that the variations in the life cycles of the plants from these three localities may be due to genetic differences because, although the chromosome numbers may be identical, this would not exclude such a possibility. This extremely large genus is now subdivided, and recently a number of new genera have been established (Hamel, 1939).

*ECTOCARPACEAE: *Pylaiella* (after de la Pylaie, a French botanist). Figs. 91, 92.

There is only one species, *P. littoralis*, and although it is said to possess a number of varieties yet it is by no means certain that they may not be ecological or seasonal forms because it has been shown that the movement of the water can even affect the nature of the branches. In general appearance the plants are very like *Ecto-*

carpus, and for many years the species was included in that genus. The branching is opposite or alternate, but the branches do not end in a mucilage hair as they do in *Ectocarpus*. Attachment to the host plants or to the substrate is by means of rhizoidal filaments, and near the base the main filaments of the erect thallus are frequently coalesced into a rope-like structure as a result of wave action. In some places the plants appear to be confined principally

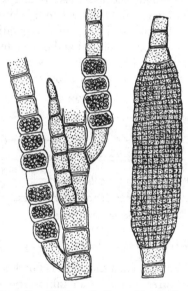

Fig. 91. *Pylaiella littoralis*. Portion of plant with plurilocular and unilocular sporangia (× 200). (Original.)

to certain host plants whilst in other areas there may be no special hosts. In the Isle of Man Knight (1923) has shown that in the spring the plants occur on *Ascophyllum nodosum*, in early summer they are to be found on *Fucus vesiculosus* and in late summer on *F. serratus*, yet in north Norfolk the species frequently grows on the stable mud banks of salt marsh creeks or else on *F. vesiculosus*. On the Swedish coast three forms have been noted, two of which are found on *Ascophyllum nodosum*, whilst the third, which is a vernal form that dies off at the end of June, occurs attached to stones. Of the two forms observed on *Ascophyllum* it is found that those directly attached to the host are the more numerous, and although they persist for the whole year they are most fertile in winter when they produce unilocular sporangia. The other plants are really epizoic because they grow on the colonies of *Sertularia* (a hydroid) that are to be found on the *Ascophyllum*. These plants, which only bear plurilocular sporangia, are most vigorous during spring and early summer and are dead by the end of July.

This species is readily distinguished from *Ectocarpus* by the position of the sporangia because these bodies are nearly always intercalary, very rarely terminal, and when this latter is the case it is frequently due to the loss of the terminal vegetative portion. The unilocular sporangia are cask-shaped and open laterally, dehiscence of the sporangium being brought about by the swelling up of the middle layer of the wall, but this process is dependent on the

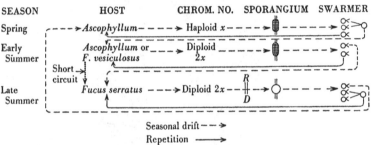

Fig. 92. *Pylaiella littoralis*. The life cycle according to Knight.

temperature of the water when the plant is flooded by the incoming tide, high temperatures acting in an inhibitory manner. Meiosis takes place in the unilocular sporangia, and each zoospore when it finally emerges possesses one nucleus, two plastids and flagellae and one eye-spot. After emergence the zoospores usually germinate singly but they have been known to fuse and thus restore the diploid condition. The plurilocular sporangia, which are produced on haploid or diploid plants, are oblong or irregularly cylindrical and also dehisce laterally, each cell producing one zooid which emerges singly. The zooids from these sporangia either fuse or else develop at once, the parthenogenetic zooids arising from diploid sporangia, principally during the summer in England and throughout the winter in Sweden, although isolated cases may occur at any time in the year. The other zooids, which function as gametes or which may occasionally develop parthenogenetically, arise from haploid sporangia and are most abundant in spring and early summer. Fig. 92 is a schema taken from Knight (1923) to illustrate the life cycle as found in English plants during the course of one year.

ECTOCARPACEAE: *Phaeostroma* (*phaeo*, brown; *stroma*, mattress).
    Fig. 93.

This is cited as an example of a much reduced ectocarpoid form which occurs as an epiphyte or partial parasite upon marine grasses, such as *Zostera*, or else upon other brown algae.

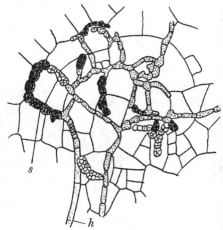

Fig. 93. *Phaeostroma Bertholdi.* Thallus ramifying in *Scytosiphon* showing sporangia (*s*) and a hair (*h*). (After Oltmanns.)

MESOGLOIACEAE: *Mesogloia* (*meso*, middle; *gloia*, slime). Fig. 94.

In this and related genera (*Castagnea, Eudesme, Chordaria* and *Acrothrix*) the construction is of the "cable" or consolidated type described by Church (1920) in which there are one or more erect parallel strands enclosed in a mucous matrix, the whole being interwoven with lateral branches. There are three principal zones that can be recognized in the plant thallus:

(*a*) a *medulla* composed either of one long thread accompanied by offshoots of the first order or else of a group of long threads;

(*b*) a *cortex* of peripheral assimilatory filaments and colourless hairs;

(*c*) a *subcortex* composed of offshoots from the medulla.

There is also a certain amount of secondary tissue which in some parts may be rhizoidal in character.

In *Mesogloia* there is a *single* central strand terminating in a hair and having a distinct intercalary meristem just below the apex.

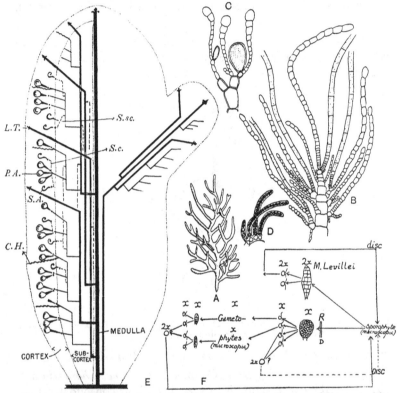

Fig. 94. *Mesogloia vermiculata.* A, plant (sporophyte). B, apex of filament with branches and beginning of cortication (× 135). C, unilocular sporangia. D, plurilocular gametangia on gametophyte. E, diagram to illustrate construction of thallus (central thread type). *C.H.* = colourless hair, *L.T.* = leading thread with intercalary growth zone, *P.A.* = primary assimilator, *S.A.* = secondary assimilator, *S.c.* = secondary cortex, *S.sc.* = secondary sub-cortex. F, diagram to illustrate life cycle. (A, C, D, after Tilden; B, E, F, after Parke.)

The cortex is formed of short horizontal filaments with somewhat globose terminal cells that are packed in a gelatinous material. The hairs, which are frequently worn away in the older parts of the thallus, occupy a *lateral* position, but owing to inequalities of

growth they may appear to be terminal. The unilocular sporangia
are ovoid and are borne at the base of the cortical filaments, but the
elongate plurilocular sporangia, which incidentally are only known
for *M. Levillei*, replace the terminal portion of the assimilatory
hairs and hence are always stalked. Meiosis takes place in the
unilocular sporangia during zoospore formation, and culture
experiments on *M. vermiculata* carried out by Parke (1933) have
demonstrated conclusively that the adult macroscopic plant of
summer and autumn is diploid, the zooids from the unilocular
sporangia germinating into a minute winter gametophyte (haploid
adelophycée form) that bears plurilocular sporangia of an ectocarpoid type. The zooids from these sporangia fuse and the zygote
develops into the characteristic basal disk from which the central
erect filament of the macroscopic plant arises. There is thus
an alternation of morphologically distinct generations in this
species. The fate of the zooids from the plurilocular sporangia
of *M. Levillei* is not known.

Mesogloiaceae: *Eudesme* (well-binding). Fig. 95.

E. virescens, which is the type species of this genus, has recently
been removed from the genus *Castagnea* to which it is very closely
allied in structure. The branched mucilaginous plants differ from
Mesogloia fundamentally in the presence of more than one central
strand in the medulla. The primary filaments in the medulla, which
originate from a basal disk, have an intercalary growing zone and
terminate in a colourless hair, and as branching takes place from
these primary filaments laterals may develop in such a manner as to
make it difficult to distinguish them from the primaries. The cortex
is composed of club-shaped primary and secondary assimilatory
hairs arranged either singly or in falcate tufts. The unilocular
sporangia develop as outgrowths from the basal cells of the primary
assimilatory filaments, whilst the plurilocular sporangia appear in
secund rows on the outermost cells of the same type of filament.
The zooids from the unilocular sporangia germinate immediately,
or else some considerable time may elapse, perhaps as much as
3 years according to some observers, before any development takes
place. They give rise to a microscopic plethysmothallus on which
plurilocular gametangia similar to those of *Mesogloia* are to be
found. After zooids have been liberated from the plurilocular

sporangia of the plethysmothallus young macroscopic *Eudesme* plants appear, so that it may be assumed that there is a definite alternation of generations in which the small gametophyte forms the winter phase.

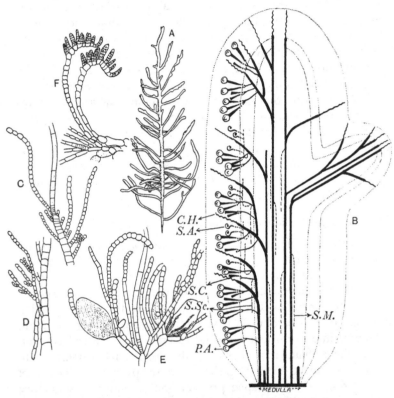

Fig. 95. *Eudesme virescens*. A, plant. B, diagram to illustrate thallus construction (multiple strand type). *C.H.* = colourless hair, *P.A.* = primary assimilator, *S.A.* = secondary assimilator, *S.C.* = secondary cortex, *S.Sc.* = secondary subcortex, *S.M.* = secondary medulla. C, apex (× 160). D, thallus with branch and corticating filament (× 75). E, unilocular sporangia. F, plurilocular sporangia. (A, E, F, after Oltmanns; B–D, after Parke.)

MESOGLOIACEAE: *Chordaria* (a small cord). Fig. 96.

In *Mesogloia* there is a single central filament whilst in *Eudesme* there are several, but in *Chordaria* development has proceeded a stage farther and the branched cartilaginous fronds possess a firm, pseudo-parenchymatous medulla of closely packed cells that have become elongated in a longitudinal direction. The cortex is composed of crowded, radiating, assimilatory filaments, which are either

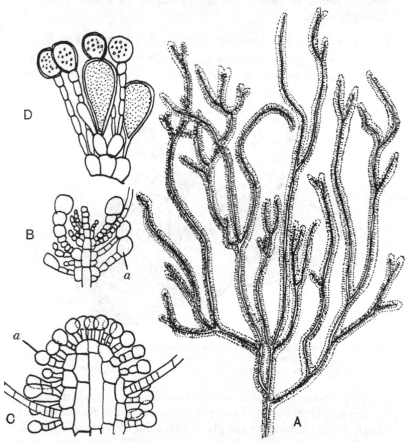

Fig. 96. *Chordaria divaricata.* A, plant (× ⅔). B, apex of young plants showing commencement of cortication. C, apex of older plant of *C. flagelliformis* showing structure of thallus. *a* = assimilator. D, unilocular sporangia (× 300). (A, D, after Newton; B, C, after Oltmanns.)

simple or branched, the whole being embedded in a thick layer of jelly, thus giving the plant a slimy touch. This type of structure, even though the growth is still confined to the apex, marks the highest development of Church's consolidated or cable type of construction. The oblong unilocular sporangia are borne at the base of the assimilatory filaments, but plurilocular sporangia are unknown. When this genus comes to be investigated it will probably be found to have a life history similar to that of the other Mesogloiaceae.

CORYNOPHLOEACEAE: *Leathesia* (after G. R. Leathes). Fig. 97.

The present genus provides an example of degeneration in the cable type of construction. The young plant arises from a small,

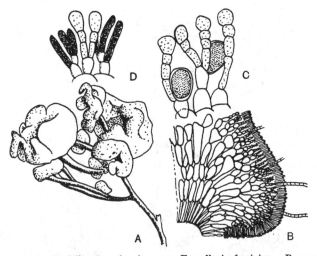

Fig. 97. *Leathesia difformis*. A, plants on *Furcellaria fastigiata*. B, transverse section to show thallus construction (×24). C, unilocular sporangia (×336). D, plurilocular sporangia (×336). (A, after Oltmanns; B–D, after Newton.)

creeping, rhizomatous portion and is composed of a packed mass of radiating, dichotomously branched filaments which are sufficiently closely entwined to make the plant mass solid. From these medullary filaments there arises a cortex of densely packed assimilatory filaments. The young plants are subspherical at first, but with increasing age the central medullary filaments commence to disintegrate and as a result the mature thallus becomes hollow and irregularly lobed. Plurilocular and unilocular sporangia are known, the zoospores from the ovoid unilocular sporangia germinating to disk-like plantlets on which plurilocular gametangia ultimately appear. These plantlets either give rise to other similar plantlets or else to the adult thallus once more. By analogy with other species the dwarf plantlets with the plurilocular sporangia may be regarded as haploid gametophytes.

ELACHISTACEAE: *Elachista* (*elachistos*, very small). Fig. 98.

Church (1920) regarded this genus as being explicable morphologically on the cable type of construction, although it must be

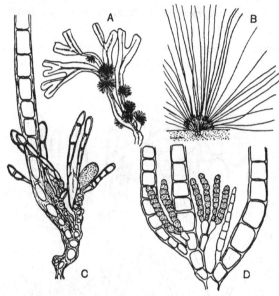

Fig. 98. *Elachista fucicola*. A, plants on *Fucus vesiculosus* (× 0·36). B, single plant in section showing penetrating base, crowded sporangia, short paraphyses and long assimilators. C, unilocular sporangia (× 120). D, plurilocular sporangia at base of assimilation thread (× 220). (A–C, after Taylor; D, after Kylin.)

regarded as a degenerate type in which the true structure is only seen in the sporeling. This possesses a horizontal portion from which a number of erect filaments arise, so that in the early stage it is comparable morphologically to *Eudesme*. In the older plant the erect filaments have developed to form a cushion composed of densely branched filaments matted together and only becoming free at the surface. The various species are epiphytic on other algae, *Elachista fucicola* being especially abundant on species of *Fucus*. The pluri- and unilocular sporangia, together with the long hairs, arise from the base of the short filaments or paraphyses. The zooids from the unilocular sporangia germinate in late autumn to give a branched, thread-like, microscopic gametophyte which persists throughout the winter. In late winter and spring plurilocular sporangia develop on the minute gametophytes, and when the zooids have been liberated they fuse and the zygote germinates into a new macroscopic *Elachista* plant.

MYRIONEMACEAE: *Myrionema* (*myrio*, numerous; *nema*, thread). Fig. 99.

The various species are epiphytic upon other algae, forming thin expansions or minute flattened cushions or disks that are very

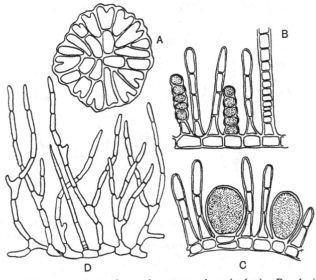

Fig. 99. *Myrionema strangulans*. A, young plant (×640). B, plurilocular sporangia (×340). C, unilocular sporangia (×340). D, 11-day-old plant from zoid of unilocular sporangium (×336). (After Kylin.)

variable in shape and from which numerous, closely packed, erect filaments and hairs arise. *M. strangulans* is especially common on sheets of *Ulva* during the summer. The basal portion of the thallus has a marginal growing region and is composed of crowded radiating filaments that may, on rare occasions, penetrate the host plant. The unilocular sporangia, which are not borne on the same

plants with plurilocular sporangia, give rise to haploid zooids, and
these develop into a thread-like gametophytic plant bearing long
filaments, the possession of this type of gametophyte indicating
that the genus is perhaps more closely allied to the Mesogloiaceae
than to the Ectocarpaceae.

SPERMATOCHNACEAE: *Spermatochnus* (*sperma*, seed; *tochnus*, fine
down). Fig. 100.

This is essentially one of the corticated types, the filamentous,
cylindrical, branched thallus being derived primarily from a central

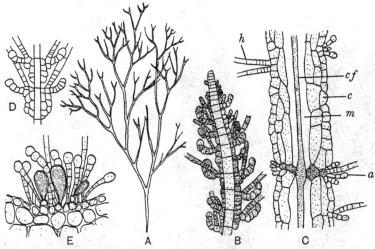

Fig. 100. *Spermatochnus paradoxus.* A, plant (×0·44). B, apex of young plant
showing origin of cortication. C, portion of old thallus showing structure.
a = assimilator, *c* = cortical cells, *cf* = central filament, *h* = hair, *m* = mucilage.
D, portion of thallus showing cortication and pairs. E, paraphyses and unilocular
sporangia (×200). (A, E, after Newton; B–D, after Oltmanns.)

axis composed of a single filament with a definite apical cell. Each
individual cell of this filament segments at one end and so definite
nodes are formed. The corticating filaments arise from the nodes,
and growth of the cortex is secured by tangential division of the
primary corticating cells, though later more filaments may grow on
top of them. The outermost layer of the cortex bears the assimilatory filaments and hairs. As the plants become older mucilage
develops internally and forces the cortex away from the primary
central filament although a connexion is maintained by the threads

from each node. Unilocular sporangia, together with clavate paraphyses, develop in sori, the sporangia arising from the base of the sterile threads. The life cycle has not yet been worked out, but if it is at all comparable with the other closely related genera then the zooids should give rise to a microscopic gametophyte generation.

SPOROCHNACEAE: *Sporochnus* (*sporo*, seed; *chnus*, wool). Fig. 101.

The thallus, which is composed of an inner layer of large cells with an outer layer of small assimilatory cells, is filamentous with

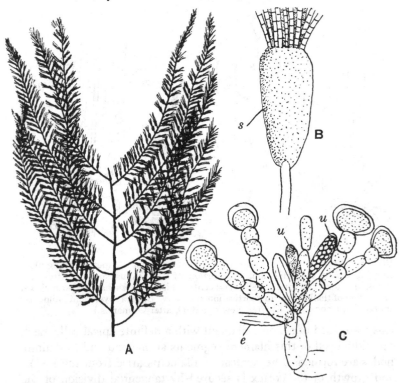

Fig. 101. *Sporochnus pedunculatus*. A, plant. B, fertile branch with receptacle. *s*=sorus. C, unilocular sporangia (*u*). *e*=empty sporangium. (After Oltmanns.)

branches arising alternately and arranged in one plane. On account of its structure Church considered that it really belonged to the parenchymatous type of construction, and morphologically it

may represent a transition stage from the corticated to the parenchymatous type, although it is usually considered that these two
forms of thallus arose independently. The unilocular sporangia,
which are club-shaped, are borne on branched monosiphonous
filaments crowded together in large oval or elongate receptacles
that bear a cluster of hairs at their apex. This type of reproductive
structure is more or less unique among the simpler forms of the
Phaeophyceae.

SCYTOSIPHONACEAE: *Phyllitis* (*phyllos*, leaf). Fig. 102.

The unbranched fronds are expanded, membranous, leaf-like
structures with an internal medulla composed of large, colourless

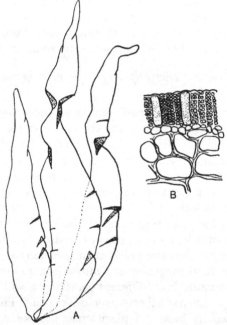

Fig. 102. *Phyllitis Fascia.* A, plant (× ⅔). B, transverse section of thallus with
plurilocular sporangia (× 375). (A, original; B, after Setchell and Gardner.)

cells and an outer layer of small, superficial, assimilatory cells.
Unilocular sporangia are not known nor are there any paraphyses.
The plurilocular sporangia, which are arranged at right angles to
the surface, arise from the superficial cells and produce zooids that

germinate to give a creeping basal thallus from which a new plant arises. It is therefore suggested that the plants are wholly diploid and that the haploid generation has been lost. Yendo (1919), however, has reported that these zooids can develop after a resting period into minute protonemal threads bearing antheridia and oogonia which presumably produced gametes, although no sign of fertilization was observed. If these observations are correct this genus must be regarded as anomalous, because normally the gametophytic generation arises from the products of unilocular sporangia. It would therefore seem premature to accept this peculiar life cycle without further evidence, and at present it would be more in agreement with known life cycles if the plants are simply regarded as being wholly diploid and without a haploid generation.

In the related genus *Scytosiphon* it would also seem that only the diploid generation is present and that the reported protonemata are not gametophytic as has been suggested by some workers.

DICTYOSIPHONACEAE: *Dictyosiphon* (*dictyo*, net; *siphon*, tube). Figs. 103, 105.

The filamentous plants arise from small lobed disks and have either a few or many branches, the younger ones commonly being clothed with delicate hairs. There is a central medulla of large elongated cells and a cortex of small cells, but in old plants the medulla is often ruptured and the axis becomes partially hollow. On the macroscopic plants only unilocular sporangia are found, each of which is formed from a single subcortical cell. Meiosis takes place in these sporangia and the zooids germinate to form microscopic prothalli: these represent the gametophytic generation and reproduce by means of plurilocular gametangia. The gametes either develop parthenogenetically into a new protonema or else two of them, coming from different gametangia, will fuse and the zygote develops into a small ectocarpoid-like plant. This may either reproduce itself by means of plurilocular zoosporangia or else it develops into a plantule from which the adult sporophyte arises.

ASPEROCOCCACEAE: *Asperococcus* (*aspero*, rough; *coccus*, berry). Figs. 104, 105.

The structure of the adult plant is essentially the same as that of the two preceding genera except that the central filaments

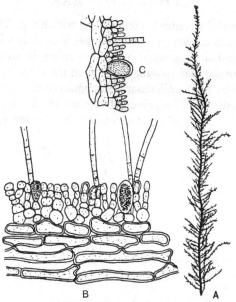

Fig. 103. *Dictyosiphon.* A, plant. B, portion of thallus of the closely related genus *Gobia* showing mode of construction with hairs, unilocular sporangia and assimilators. C, unilocular sporangia of *Gobia*. (After Oltmanns.)

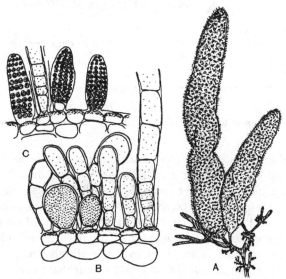

Fig. 104. *Asperococcus bullosus.* A, plant. B, unilocular sporangia (×225). C, plurilocular sporangia (×225). (A, after Oltmanns; B, C, after Newton.)

degenerate and the centre becomes filled with a gas. The fronds are
simple or branched and bear small superficial cells with sporangia
and mucilage hairs scattered over the surface in sori. The plurilocular and unilocular sporangia occur on the same or on different
plants, the sori with unilocular sporangia containing sterile paraphyses in addition. The principal interest of this type is centred
around its life history which has been studied by several workers in

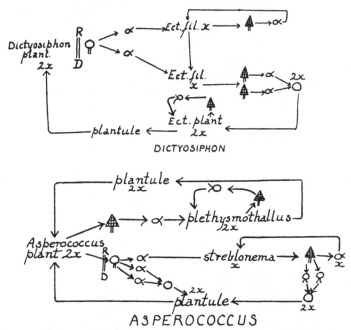

Fig. 105. Diagram of life cycles of *Dictyosiphon foeniculaceus* and *Asperococcus
bullosus*.

considerable detail. In *A. compressus* the life cycle is simple, the
zooids from the unilocular sporangia germinating directly into a
protonemal phase that later turns into small plantules; these can
reproduce themselves successively by means of zoospores from
both pluri- and unilocular sporangia until the advent of favourable
conditions enables the development of the macroscopic phase to
take place once more. There is no evidence of either meiosis or of
gametic fusion. In *A. fistulosus* it would appear that the life cycle is
dependent upon the behaviour of the zooids from the unilocular

sporangia where meiosis has been shown to take place. If they fuse, the zygote develops first into a "*streblonema*" phase, so-called from the brown alga it resembles, and then into a plantule from which a new adult plant can arise. In this case there is no evidence for the existence of a gametophytic generation, nor has any evidence been obtained to show that such streblonemoid plants can reproduce themselves by means of sporangia. If no fusion of the zooids from the unilocular sporangia takes place the "*streblonema*" phase is again produced parthenogenetically, but under these circumstances plurilocular sporangia are formed which give rise to

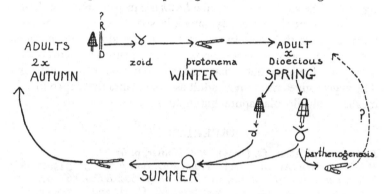

COLPOMENIA SINUOSA

Fig. 106. Life cycle of *Colpomenia sinuosa*. RD = probable place of reduction division in life cycle. (Modified from Kunieda and Suto.)

a new "*streblonema*" generation, nor has any investigator under such conditions succeeded in obtaining macroscopic plants again and so it has been suggested, therefore, that sex has been inhibited in these plants. In *A. bullosus* the zooids from the plurilocular sporangia on the macroscopic thallus do not fuse but germinate directly to give rise to a series of plethysmothalli bearing plurilocular sporangia: these tide over the winter season, and then in spring young *Asperococcus* plants develop in place of the sporangia on the ectocarpoid plantules. In the unilocular sporangia meiosis takes place and the zooids develop into minute gametophytic plants that produce plurilocular sporangia. If the gametes from these sporangia fuse the zygote develops into a plantule from which a new macroscopic plant arises, but if there is no fusion then they merely develop into a new gametophytic generation. Sauvageau also reported that the zooids from the unilocular sporangia may give rise to creeping filaments which later produce young plantules of *Asperococcus*. This direct reproduction of the macroscopic plants can only be explained by a premature abnormal fusion of some of the zooids from the unilocular sporangia. It will be evident that direct alternation of generations is obscured in this type through the number of possible independent circuits and "short cuts". Recently the life cycle in *Colpomenia sinuosa*, a member of a closely allied genus, has been described in detail (cf. fig. 106). The adult plants are like *Leathesia* in appearance, although they are essentially parenchymatous in structure. It will be seen, however, that there are two morphologically similar generations, the dioecious gametophytes appearing in spring and reproducing by means of anisogametes that are formed in dissimilar gametangia. The zygote gives rise to new adult asexual plants that reproduce by means of plurilocular sporangia in autumn.

REFERENCES

Ectocarpus. HAMEL, G. (1939). *Bot. Notiser*, p. 65.

Pylaiella. KNIGHT, M. (1923). *Trans. Roy. Soc. Edinb.* **53**, 343.

Ectocarpus. KNIGHT, M. (1929). *Trans. Roy. Soc. Edinb.* **56**, 307.

Asperococcus. KNIGHT, M., BLACKLER, M. C. H. and PARKE, M. W. (1935). *Trans. Lpool Biol. Soc.* p. 79.

Colpomenia. KUNIEDA, H. and SUTO, S. (1938). *Bot. Mag., Tokyo*, **52**, 539.

General. KYLIN, H. (1933). *Lunds Univ. Årsskr.* N.F. Avd. 2, **29**, no. 7, p. 1.

Ectocarpus. PAPENFUSS, G. (1935). *Bot. Gaz.* **96**, 421.

Mesogloia, Castagnea, Acrothrix. PARKE, M. W. (1933). *Publ. Hart. Bot. Lab.* no. 9.

Ectocarpus. SCHUSSNIG, B. and KOTHBAUER, E. (1934). *Öst. Bot. Z.* **83**, 81.

Phyllitis. YENDO, K. (1919). *Bot. Mag., Tokyo*, **33**, 171.

CUTLERIALES (ISOGENERATAE)

This order is characterized by trichothallic growth, *regular* alternation of generations, and a well-marked anisogamy which in some respects approaches oogamy. They are generally placed in the Isogeneratae, even though this leads to a difficulty because in *Cutleria* the two generations are not equal morphologically although they are equal in *Zanardinia*. On this classification, therefore, *Cutleria* must be regarded as a modified member of the Isogeneratae or else it must be separated from *Zanardinia* and put in a separate family in the Heterogeneratae.

CUTLERIACEAE: *Cutleria* (after Miss Cutler). Fig. 107.

The gametophyte and sporophyte generations are distinctly heteromorphic and also differ in their seasonal occurrence, the former being a summer annual whilst the latter is a perennial reaching its maximum vegetative phase in October and November

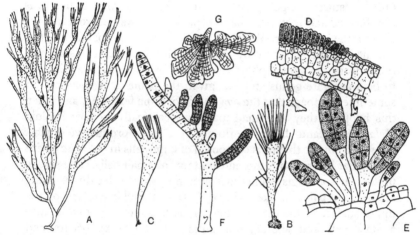

Fig. 107. *Cutleria multifida.* A, plant (× ⅓). B, young seedling. C, seedling slightly older to show branching. D, transverse section of thallus with unilocular sporangia. E, female gametangia. F, male gametangia. G, "*Aglaozonia*" stage. (A, original; B–D, G, after Oltmanns; E, F, after Yamanouchi.)

with a peak fruiting period in March and April. The gametophyte is an erect flattened thallus rendered fan-like because of the repeated dichotomies. The thallus and apices are clothed with tufts of hairs, each with a basal growing region, whilst the oogonia and antheridia, which are borne on separate plants, occur in sori on both sides of the thallus. The antheridia, with which hairs are sometimes associated, are formed in clusters from superficial cells of the thallus that divide to produce a stalk cell and an antheridium initial. The mature antheridium contains about 200 antherozoids, each of which possesses two chromatophores, and they are much smaller than the mature ova, each of which contains thirty or more chromatophores.

The oogonia, with which hairs are sometimes associated, are also formed from superficial cells which divide into a stalk cell and an oogonium initial. The ripe oogonium contains sixteen to fifty-six eggs which, after liberation, remain motile for a period of from 5 min. to 2 hours, whilst the antherozoids can remain active for about 20 hours. Discharge of the gametes takes place at any time during the day but is at its best about 5 a.m., fertilization taking place in the water when the diploid number of chromosomes (48) is restored. Upon germination a small columnar structure is first formed and then a flat basal expansion grows out from its base to form the adult sporophyte, which is a prostrate expanded thallus attached to the substrate by means of rhizoids. It differs so very much from the gametophyte that when first found it was thought to be a separate genus and was given the name of *Aglaozonia*. It sometimes happens that the ova do not become fertilized, and when this happens they germinate parthenogenetically to give haploid *Aglaozonia* plants, but these do not bear any reproductive organs. The sporophytic thallus is composed of large cells in the centre with superficial layers, both top and bottom, of small cells. The sessile unilocular sporangia, sometimes accompanied by deciduous hairs, are borne in palisade-like sori or else are scattered irregularly on the upper surface of the thallus. Each superficial cell first divides into a stalk cell and sporangium initial, then meiosis occurs and eventually eight to thirty-two zoospores are formed in each sporangium. The zoospores on germination give rise to new *Cutleria* plants. This life cycle was first worked out by Yamanouchi (1912) for the common species *Cutleria multifida* and its sporophyte *Aglaozonia reptans*.

SPHACELARIALES (ISOGENERATAE)

The next three types belong to the Sphacelariales, an order frequently known as the "Brenntalgen" because they possess a very characteristic large apical cell with dense brown contents, the detailed classification of the group being based primarily upon the behaviour of this apical cell at branch formation. The plants have regular branching and a bilateral symmetry, both of which form characteristic features. Structurally they can be regarded as strengthened multiseriate filaments, *Sphacella* perhaps being one of the more primitive members of the group with a non-corticate monosiphonous axis.

*SPHACELARIACEAE: *Sphacelaria* (gangrene). Fig. 108.

The plants are filamentous with hypacroblastic branching in which the cell below the apex cuts off two branch initials opposite to

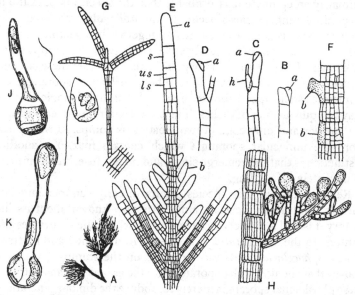

Fig. 108. *Sphacelaria*. A, plant of *S. cirrhosa* (× ½). B, apical cell (*a*) of *S. cirrhosa*. C, *S. cirrhosa*, origin of hair (*h*). *a* = apical cell. D, hair at older stage; *a* = apical cell. E, apex of thallus of *S. plumigera* showing branches, *b*; single segment (*s*), which later divides into upper (*us*) and lower (*ls*) segments; *a* = apical cell. F, origin of branch, *b*. G, bulbil of *S. cirrhosa* (× 52·5). H, unilocular sporangia, *S. racemosa*. I, zoospore of *S. bipinnata* (× 1200). J, K, germinating spore of *S. bipinnata* (× 1200). (A, original; B–G, after Oltmanns; H, after Taylor; I–K, after Papenfuss.)

each other, although in some cases the initials may remain dormant. The plants grow attached to stones or other algae by means of basal disks or rhizoids that have spread down from the lower cells of the stalk. Mucilage hairs, which arise from the apical cell through a segment being cut off obliquely, are present in some species though they may disappear with age. The axis and main branches form a solid frond due to cortication which commences near the apex through a series of transverse and longitudinal divisions, until finally there is an external layer of rectangular cells arranged in a polysiphonous manner. Unilocular and plurilocular sporangia are formed on short pedicels, usually on separate plants.

Clint (1927) has studied in some detail the life cycle of *S. bipinnata*, which grows epiphytically on *Halidrys* (cf. p. 203) in the Irish Sea, whilst farther south it frequents *Cystoseira* (cf. p. 205). Although primarily an epiphyte it is probable that the species is parasitic to a certain extent. Meiosis occurs in the unilocular sporangia, and after the zooids have been ejected all together in a gelatinous mass they fuse in clumps of two to five, the cytology of the clumps being unknown. Isolated spores may germinate, but under these circumstances the sporeling soon dies. The plurilocular sporangia are sometimes stalked, and so it is suggested that morphologically they may be equivalent to branches. The zoospores from these sporangia are smaller and only contain two plastids as compared with those from the unilocular sporangia which contain four. Yet another distinction is that they emerge singly and do not fuse, but germinate immediately on settling.

Reproduction in this species occurs in early summer and late autumn, and whilst a certain amount is now known about its life history it is still a mystery as to how or in what state it survives the winter. In the north *Halidrys* breaks off in the winter and no trace of any *Sphacelaria* plants can be found on the stumps. It is now known that the unilocular sporangia are the asexual organs and that these plurilocular sporangia merely reproduce the diploid generation. The morphologically similar gametophyte generation has since been found and it gives rise to isogametes from plurilocular sporangia. There is thus a regular alternation of generations which agrees with the facts for other members of the family. Vegetative reproduction also takes place in this genus by means of modified branches or propagules which are usually pedicellate and triradiate, the actual shape varying for the different species. The tropical species, *S. tribuloides*, is said to form the common food of many Hawaian fishes.

CLADOSTEPHACEAE: *Cladostephus* (*clado*, shoot; *stephus*, a crown). Fig. 109.

The plants, which are bushy in appearance, arise from well-developed holdfasts and are characterized by the ecorticate branches being arranged in whorls with tufts of hairs just below their apices. Cells just below the apex divide to give a number of branch segments, this type of branching being known as *polyblastic*. The main axis is corticate and primarily polysiphonous, because the subterminal cells divide to form cortical cells which then divide again several times, but as there is also an outer cortex of rhizoids a pseudo-parenchymatous structure is ultimately formed. Both unilocular and plurilocular sporangia are formed on *stichidia* which

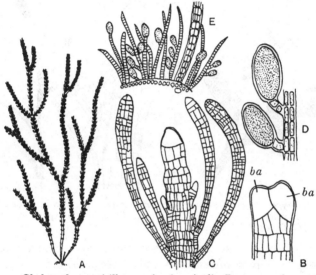

Fig. 109. *Cladostephus verticillatus*.   A, plant (×½).   B, apex to show origin of branch.   *ba* = branch apex.   C, thallus showing cortication.   D, unilocular sporangia (×225).   E, part of thallus with unilocular sporangia (×45).   (A, D, E after Newton; B, C, after Oltmanns.)

arise from the rhizoidal cortex in the internodes between the whorls of vegetative branches. The two different types of sporangia occur on separate plants, the unilocular on what must be the diploid generation as they produce zoospores, and the plurilocular on what must be the haploid generation because they produce isogametes.

STYPOCAULACEAE: *Stypocaulon* (*stypo*, coarse part of flax; *caulon*, stem). Fig. 110.

The pinnate frond arises from a well-marked basal system, the plants in summer having the appearance of shaggy tufts whilst in

winter they are more regularly pinnate owing to the shedding of
branches. The inner cortex of the central axis is composed of a
number of cubical cells whilst there is also an outer cortex of
rhizoidal cells, the whole forming a pseudo-parenchyma. Any cell,
whether in the inner or outer cortex, can develop a new apical cell
upon injury, so that there is a great power of regeneration of apical

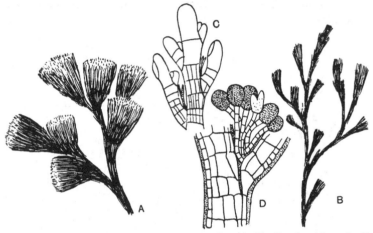

Fig. 110. *Stypocaulon scoparium.* A, summer form (× ¾). B, winter form (× ¾).
C, apex, showing branches with cortication. D, unilocular sporangia. (A, B,
original; C, D, after Oltmanns.)

cells. Only unilocular sporangia are known, and these are formed
in groups of up to fifteen on a pad of tissue in the axil of each
branch on the fertile shoot, but as in *Cladostephus* secondary
sporangia may arise within the empty sheaths of the old ones.
Meiosis takes place in these sporangia, and the zooids on germination give rise to new *Stypocaulon* plants. One herbarium plant of
S. scoparium with antheridia and oogonia has been reported, but
these may have been abnormal unilocular sporangia. In view of the
great interest of this observation, however, it would be very
desirable to have a further study made of this alga.

TILOPTERIDALES (ISOGENERATAE)

Haplospora (*haplo*, simple; *spora*, seed). Fig. 111.

This and *Acinetospora* belong to a peculiar group of algae, the
life cycles of which are somewhat incompletely known, but it is

possible that they represent a transition towards the tetrasporic
Dictyotales (cf. p. 163). The chief characteristic of the asexual plant
is reproduction by means of quadrinucleate monospores which may
be equivalent to unsegmented or primitive tetraspores, although
they might equally well be degenerate tetraspores. Sexual reproduction is brought about by means of gametes from microgametangia
and larger associated sporangia that may represent oogonia, but

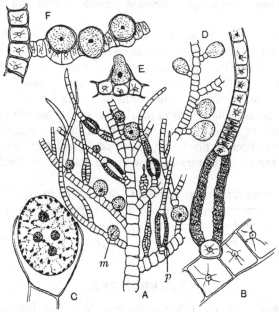

Fig. 111. *Haplospora globosa.* A, portion of plant with uninucleate sporangia,
m (oogonia?), and plurilocular microgametangia, *p*. B, plurilocular micro-
gametangium. C, monosporangium with quadrinucleate monospore. D, monosporangia. E, F, unilocular sporangia (oogonia?). (A–C, after Oltmanns, D–F,
after Tilden.)

as fertilization has not been observed there is an opportunity here
for future research which should also determine whether there
is a regular alternation of generations. The evidence at present
available suggests that there is probably a regular alternation of two
similar generations.

The sexual plants develop intercalary tubular microgametangia
which are produced by the transformation of one or more cells of
the main filament. Besides these organs there are the larger and

spherical uninucleate sporangia (oogonia?) borne on a stalk cell and partly immersed in the branches. The asexual plant reproduces by means of quadrinucleate spores formed singly in stalked or sessile, terminal or intercalary, monosporangia. Meiosis has been reported as occurring in these sporangia and this would be expected if they were primitive tetraspores. It would appear, according to some accounts, that the plants known as *Haplospora globosa* and *Scaphospora speciosa* are simply alternate phases of one and the same species.

In *Acinetospora* the plant structure is very simple, the slender, tufted thallus being monosiphonous throughout and frequently unbranched or else with very occasional branches. No fusion of zooids from either the uni- or plurilocular sporangia has been observed, and so this alga must either be regarded as the simplest member of the Tilopteridales in which sexuality has not yet wholly developed, or else as a degenerate member in which the sexual organs have been lost or highly modified. This latter view is probably the more satisfactory in view of the position of the family as a whole.

Plants with unilocular sporangia only occur in April and May and the swarmers give rise to plants bearing plurilocular sporangia. It has recently been suggested that the monospores are a means of vegetative reproduction, e.g. morphologically equivalent to propagules.

REFERENCES

Sphacelaria. CLINT, H. B. (1927). *Publ. Hart. Bot. Lab.* no. 3, p. 1.
Stypocaulon. HIGGINS, E. M. (1931). *Ann. Bot., Lond.*, **45**, 345.
Acinetospora. SCHMIDT, P. (1940). *Ber. dtsch. Gesell.* **58**, 23.
Cutleria. YAMANOUCHI, S. (1912). *Bot. Gaz.* **54**, 441.

CHAPTER VII

PHAEOPHYCEAE (CONT.)

DICTYOTALES, LAMINARIALES AND FUCALES

DICTYOTALES (ISOGENERATAE)

*DICTYOTACEAE: *Dictyota* (like a mat). Fig. 112.

This genus is representative of the Dictyotales, an order characterized by a well-marked regular alternation of two identical genera-

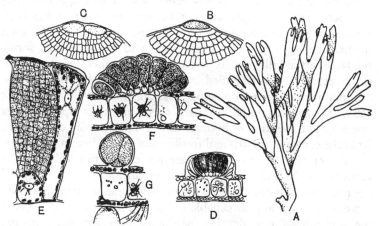

Fig. 112. *Dictyota dichotoma.* A, portion of plant showing regular dichotomy.
B, apical cell. C, apical cell divided. D, group of antheridia surrounded by
sterile cells. E, single antheridial cell and a sterile cell. F, sorus of oogonia.
G, tetrasporangium. (A–D, F, G, after Oltmanns; E, after Williams.)

tions. Asexual reproduction is brought about by means of tetraspores produced in superficial tetrasporangia, whilst the sex
organs, which are heteromorphic, are always borne in sori. The
thallus possesses a specialized bilaterality with well-marked apical
growth.

In *Dictyota*, as represented by the cosmopolitan species *D. dichotoma*, the flattened thallus exhibits what is practically a perfect
dichotomy because there is always a median septation of the apical
cell. Viewed in transverse section the thallus is seen to be composed
of three layers, a central one of large cells and an upper and lower

epidermis of small assimilatory cells from which groups of mucilage hairs arise.

The male and female sex organs are borne in sori on separate plants, the male sorus being composed of as many as 300 pluri-locular antheridia surrounded by an outer zone of sterile cells. At the formation of an antheridium a superficial cell divides into a stalk cell and an antheridium initial, the final partition of the antheridium initial into the individual antheridial mother cells taking place only a few days before the antherozoids are to be liberated. The mature antherozoid is pear-shaped with only one cilium, and as each plurilocular antheridium liberates about 1500 antherozoids, a single sorus may generate as many as 450,000. The number of ova produced are not so numerous, and it has been estimated that there are about 6000 antherozoids available for each ovum. The oogonial sorus is very similar to the antheridial sorus, the large fertile cells, twenty-five to fifty in number, being situated in the centre and surrounded by sterile cells on the outside. The oogonia likewise arise from superficial cells that divide into a stalk cell and oogonium initial, and each oogonium when ripe produces one ovum. Liberation of both kinds of gamete usually commences from the centre of a sorus and fertilization takes place in the water, but during the process the eggs are not caused to revolve by the activities of the antherozoids as they are in *Fucus* (cf. p. 197). If the process is followed under a microscope it can be noted that only some of the eggs appear capable of attracting antherozoids, whilst the unfertilized ova develop parthenogenetically; such plants, however, always die in culture, though it is possible that in nature they may persist. The sex organs are produced in regular crops, the new sori appearing between the scars of the old, and when the whole of the surface has been used up the plant dies.

After fertilization the zygote develops into a morphologically similar plant which reproduces by means of *tetraspores* that are formed in tetrads in superficial sporangia. At sporangium formation an epidermal cell swells up in all directions, and after a stalk cell has been cut off the sporangium initial divides twice to give the four tetraspores, during which the thirty-two diploid chromosomes are reduced to the haploid number of sixteen. A tetraspore at the time of liberation is an elongated body and grows at once into a new sexual plant. In some cases, however, the tetrasporangium fails to divide into four spores but germinates as a whole and this phenomenon probably explains the abundance of sporophytic plants in certain localities, although the conditions that cause this abnormality have not yet been discovered. Whilst the sex organs are produced in rhythmic crops there is no such periodicity in the case of the tetraspores, and here again there is scope for further research.

In the related genus *Taonia* the asexual plant bears tetrasporangia and hairs in zonate bands across the thallus, and there is some evidence for a correlation between the tides, or perhaps the light conditions of each intertidal period, and the development of the zones. Each zone probably corresponds to a single tidal period because a plant 30 days old was found to possess sixty zones of tetrasporangia. The period between the initiation of each new crop is probably required in order that the plant may accumulate the necessary food material. In *Taonia* also the asexual plants are frequently more abundant than the sexual, but this is partly accounted for by the persistence of a sporophytic rhizoidal portion that can give rise to new plants. More commonly, however, the tetrasporangium fails to divide and the whole structure germinates before meiosis has taken place. Plants formed in this way are found to be more resistant and vigorous than the plants produced from normal tetraspores, and this may be due to the larger supply of food material available from a complete sporangium.

Three kinds of rhythmic periodicity for the sex organs of *Dictyota* have been described from different localities:

(*a*) In Wales the sori require 10 to 13 days to develop whilst in Naples 15 or 16 days are necessary, the gametes being liberated about once a fortnight in both areas.

(*b*) In North Carolina liberations occur once a month, at the alternate spring tidal cycles, although only 8 days are required for the development of the sex organs. This suggests that the plants are exhausted after each fruiting and a resting period is necessary in order to recuperate.

(*c*) In Jamaica the successive crops take a very long time to mature, e.g. very little change can be seen even after 22 days. This results in almost continuous fruiting with two successive crops overlapping. One very significant feature is that the commonest species, *D. dichotoma*, apparently behaves as described above in each of the three localities. It is, however, possible that there is a genetical distinction between the plants from the different localities and an investigation along these lines might prove very profitable.

Wherever the plants occur the bulk of the gametes (60–70%) are usually liberated in a single hour at about daybreak. On the Welsh coast the gametes are set free just after each series of high spring tides during July to October, and it has been suggested that light plays the part of the determining factor during the intertidal periods. However, when plants were removed to the laboratory it was found that the periodicity was maintained, so that it must be inherited, whilst plants from Carolina likewise retained their periodicity when transferred to the laboratory, the specimens fruiting at the same time as those living under natural conditions. The mean tidal differences vary considerably in the four localities, 11–18 ft. in England, 0·8 ft. at Naples, 3·0 ft. in North Carolina, and 0·8 ft. in Jamaica. These differences preclude either light or tidal rise from being the controlling factor because the English and Neapolitan plants behave similarly even though there is a great difference in the tides. Regularity of the tidal cycle, however, may modify the reproductive cycle, because where the tides are somewhat irregular, as in Jamaica, the reproductive rhythm is also irregular. This rhythmic behaviour is probably not due to any one factor but has been acquired over a long period of time as a response to the environment and is now inherited. The phenomenon is not confined to *Dictyota* because regular or irregular periodic cropping has been recorded for species of *Sargassum*, *Halicystis*, *Cystophyllum*, *Padina* and *Nemoderma*. Culture experiments are required in order to determine whether the habit persists in successive generations when they are grown under completely non-tidal conditions, but unfortunately *Dictyota* has not proved very amenable to cultural conditions. Finally, it can be argued that tides and light may have no control over this rhythm and that it may be associated instead with lunar periodicity, in which case even cultures will be of no avail. It has been observed that the plants in North Carolina always fruited at the time of full moon, and it is a well-known fact that a number of marine animals spawn regularly at such a period. At present the lunar explanation would appear to be the most satisfactory, but even that produces difficulties when the behaviour of the species in Jamaica is considered.

REFERENCES

Dictyota. HOYT, W. D. (1907). *Bot. Gaz.* **43**, 383.
Dictyota. HOYT, W. D. (1927). *Amer. J. Bot.* **14**, 592.
Taonia. ROBINSON, W. (1932). *Ann. Bot., Lond.,* **46**, 113.
Dictyota. WILLIAMS, J. LLOYD (1904). *Ann. Bot., Lond.,* **18**, 141, 183.
Dictyota. WILLIAMS, J. LLOYD (1905). *Ann. Bot., Lond.,* **19**, 531.

LAMINARIALES (HETEROGENERATAE)

The Laminariales form an order which is principally temperate, the bulk of the species being confined to the colder waters of the earth, and there are, in particular, a number of monotypic genera confined to the Pacific coast of North America. The presence of such genera suggests that the original centre of distribution was in the Pacific waters that surround Japan and Alaska. The thallus, representing the large conspicuous sporophytic generation, is nearly always bilaterally symmetrical with an intercalary growing zone, whilst the gametophytes are microscopic. The sporophytes reproduce by means of unilocular zoosporangia, commonly formed in sori with paraphyses, whilst the gametophytes reproduce by means of ova and antherozoids that are borne on separate plants.

*CHORDACEAE: *Chorda* (a string). Fig. 113.

The long whip-like thallus, which is clothed in summer with mucilage hairs, arises from a small basal disk with the growing region situated just above the holdfast. The hollow fronds are simple with diaphragms at intervals, the construction of the thallus being essentially that of a multiseptate cable derived from the *Mesogloia* type by further segmentation of descending hyphae to form a pseudo-parenchyma. The epidermal layer is ultimately clothed with sporangia, paraphyses and deciduous mucilage hairs, whilst the central cells become much elongated and support the filaments that go to form the diaphragm. The zoospores on germination give rise to small filamentous gametophytes, the male plants being composed of small cells, each with two to four chloroplasts, and the female of larger cells with more numerous chloroplasts. The gametangia are borne laterally or terminally on short branches, but the plants do not become fertile for at least 3 months after their formation and they usually require 6 months. After fertilization the oospore remains attached to the wall of the oogonium. The

168 PHAEOPHYCEAE

macroscopic plant is an annual, being abundant in the colder
waters of both hemispheres.

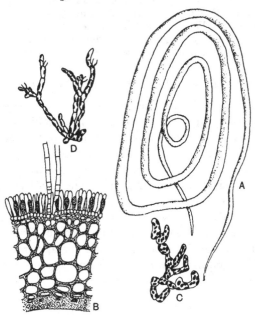

Fig. 113. *Chorda Filum.* A, plant (× ½). B, transverse section, high-power, with
sporangia. C, female gametophyte (× 145). D, male gametophyte (× 175).
(A, original; B, after Oltmanns; C, D, after Kylin.)

DESMARESTIACEAE: *Desmarestia* (after A. G. Desmarest). Fig. 114.

The plants are bushy and usually of some considerable size,
especially the species found on the Pacific coast of North America.
They sometimes bear gall-like swellings which are caused by a
copepod, and similar galls caused by the copepod *Harpacticus
chelifer* have been recorded from the red alga *Rhodymenia palmata.*
The erect, cylindrical or compressed thallus arises from a disk-like
holdfast and exhibits regular pinnate branching, the branches either
being elongate or else mere denticulations. The elongate branches
terminate in much-branched uniseriate filaments, which are also to
be found on the denticulations, but as these filaments are deciduous
the plants have a definite winter and summer aspect. Morphologically, the thallus is composed of a single prominent central row

of large cells, and these are surrounded by cortical cells which become smaller and smaller towards the periphery, the outermost layer giving rise to the branched hairs.

The unilocular sporangia are on slightly raised portions of the thallus and develop from cortical cells which undergo scarcely any modification. Meiosis takes place in the sporangium, and the ripe zoospores escape in a mass and germinate to give rise to dioecious filamentous gametophytes which are heterothallic. The smaller male plants produce terminal antheridia from each of which is

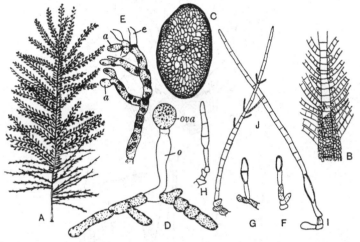

Fig. 114. *Desmarestia*. A, plant with summer and winter appearance ($\times \frac{1}{3}$). B, apex showing cortication. C, transverse section stipe. D, female gametophyte. o = oogonium. E, male gametophyte. a = antheridium, e = empty antheridium. F–J, stages in seedling germination. (A, after Newton; B, C, after Oltmanns; D–J, after Schreiber.)

liberated a single antherozoid, whilst the larger female plants produce the swollen oogonia. Each oogonium gives rise to a single ovum which escapes, but as fertilization and germination take place just outside the pore of the oogonium the young sporophyte develops as far as the monosiphonous stage whilst still possessing a primitive holdfast in the shape of the empty oogonium. Cortication, which is best observed near the apex of old plants, commences in the young plants after a few weeks, and further growth is maintained by an intercalary growing zone some way behind the apex. It is only just recently that the real life history of this genus has been established, and as a result it has seemed desirable to remove the genus from its former position in the Ectocarpales to the Laminariales.

*LAMINARIACEAE: *Laminaria* (a thin plate). Figs. 115–118.

This genus has a very wide distribution in the waters of the north temperate and Arctic zones, and it is commonly studied because its

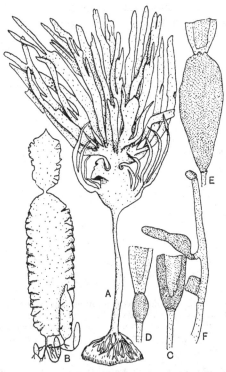

Fig. 115. *Laminaria.* A, *L. Cloustoni.* B, *L. Rodriguez.* C–E, normal regeneration ($\times \frac{1}{2}$). C, rupture just commencing. D, E, the new tissues are more heavily shaded. F, wound regeneration ($\times \frac{1}{2}$). (A, B, after Oltmanns; C–F, after Setchell.)

morphology is characteristic of the group as a whole with the exception of *Chorda* and *Desmarestia*. Furthermore, it was the first genus in which the existence of a dwarf gametophyte was established, thus leading to a new orientation of ideas in the classification of the Phaeophyceae. The expanded lamina has no mid-rib and is borne on a stipe that arises from a basal holdfast which can vary greatly in form. The simplest transition area from stipe to lamina is quite plain, but one may also find folds, ribs or callosities in that position, which is also the region of intercalary growth. *Laminaria Sinclairii* has been studied by Setchell (1905) in some detail in connexion with regeneration, a common feature throughout the genus. Three types of growth can be recognized, all of them confined to the stipe, whilst it is also possible to find all three processes taking place in one individual:

(1) The ordinary growth and extension of the blade during the growing season. This hardly merits the description of continuous physiological regeneration given to it by Setchell unless the concept of regeneration is to have a wider significance.

(2) Periodic physiological regeneration which represents the annual process whereby the new blade is formed. The transition area bulges, due to new growth in the medulla and inner cortex, and then ruptures from the pressure, thus leaving the frayed ends of the non-growing outer cortex forming collars, the upper one of which rapidly wears away. After the rupture the new cells of the medulla and inner cortex elongate rapidly. The failure of the outer cortex to grow is probably associated with the proximity of the inner cortical cells to the medullary hyphae where they can monopolize all the growing materials, thus cutting off any supply to the outer cortex, but there may, of course, be other factors involved.

(3) Restorative regeneration whereby branches arise from wounded surfaces, the same tissues being involved as in process (2) (cf. fig. 115).

Many of the species are used as food by the Russians, Chinese and Japanese. In Japan, foods derived from about ten different species of these algae are known as *Kombu*, kelp gathering from July to October forming quite a big industry. Goitre is practically unknown in Japan, and its absence must be largely connected with the iodine obtained from this algal food. Here we have an example of a region where the absence of a disease can be directly associated with the presence and nature of a particular kind of food. Apart from food the kelps are generally employed as a source of iodine and also as fertilizers.

The following brief notes concern a few species that are of more general interest:

L. Cloustoni. The attachment crampons are arranged in four lateral rows and there is a long cylindrical stipe which develops abruptly into the frond.

L. Rodriguezii. The thallus develops annually and splits near the base, the split gradually extending to the apex. Rhizoids develop on the crampons of this species.

L. saccharina. The margin is thicker than the central part of the

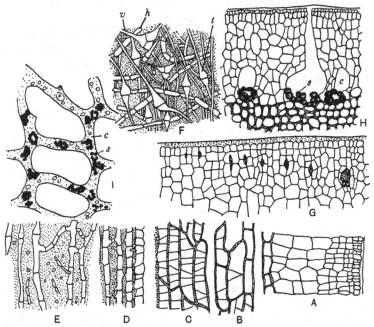

Fig. 116. Laminariaceae. A–F, portions of the stipe of *Macrocystis* passing successively from the epidermis, A, through the medulla, B–E, to the pith, F. *h* = hypha, *v* = connecting thread, *t* = "trumpet" hyphae. G, stages in development of mucilage canals, *L. Cloustoni.* H, mucilage canal of *L. Cloustoni* in transverse section. *c* = canal, *s* = secretory cells. I, mucilage canal system in *L. Cloustoni.* *c* = canal, *s* = secretory cells. (After Oltmanns.)

thallus and the wavy lamina is produced by continual growth of the central portion without any growth in the marginal areas. The stipe is short and the transition to frond is gradual.

L. digitata. This possesses a digitate frond that arises by gradual transition from a stipe which tends to be flattened, thus forming a convenient means of distinguishing it from *L. Cloustoni.*

Renfrewia. A genus very closely allied to *Laminaria* but differing from it in that there are no crampons but only a basal attachment disk.

Morphologically both lamina and stipe in *Laminaria* can be divided into three regions (cf. fig. 116), the outer cortex, the medulla, and the pith or central portion of the medulla. The one-layered blade first becomes two-layered and then the primary tubes of the medulla are cut off and separate the two outside layers.

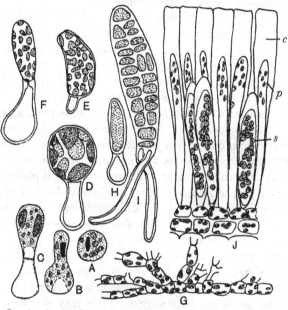

Fig. 117. *Laminaria*. A–F, stages in development of female gametophyte from a spore (A–D × 1333, E–F × 600). G, male gametophyte (× 533). H, I, first two stages in development of young sporophyte. J, sporangia (*s*), paraphyses (*p*) and mucilage caps (*c*). (A–I, after Kylin; J, after Oltmanns.)

Next the cortical cells arise from the cells of the limiting layer by divisions parallel to the surface. Somewhat later the cells of the inner cortex elongate, the middle layer of the common wall becomes swollen and the cells separate from each other except at a few points where the connexions become drawn out into short secondary tubes. Subsequent increase in thickness is due to growth in the limiting layer and the production of tubes and hyphae together with a considerable development of mucilage, so that the central cells become even more separated from each other. Two types of lateral connecting branches can be recognized, the connecting threads and hyphae. The former arise first in the course of development as outgrowths from the individual cells, but even when mature they are composed of only a few cells. The hyphae, which arise later as short branches of small cells cut off from the original vertical cells, can unite with each other or else they grow by cell division until finally they contain numerous cells which subsequently elongate very considerably.

One of the most characteristic features of the genus is the presence in the medulla of "trumpet" hyphae which are modified cells in the connecting threads and hyphae. At a transverse cell wall the ends of both cells swell out to form bulbs, the upper bulb always being larger, but so far no satisfactory explanation of this peculiarity has been advanced, though it may be due to purely mechanical requirements. The transverse wall is perforated to form a sieve plate and a callus develops on each side, both callus and sieve plate being traversed by protoplasmic strands. It will be seen that in many respects these trumpet hyphae resemble the sieve tubes of the flowering plants, but although the callus is said to be formed in land plants because of changes in pH, so far no evidence has been published to indicate whether this is also true for the Laminariaceae. Apart from the sieve plates the trumpet hyphae also possess spiral thickenings which appear as striations, and here again there is the problem of their interpretation (e.g. are they growth zones?), although it is possible that they have now lost any function they once possessed. The problem of these trumpet hyphae is still subject to considerable speculation: it has been suggested that they may be a storage or conducting tissue, whilst another suggested function is that of support, but as the plants are commonly submerged the water would seem to fulfil this requirement. In some species many of the other cells also contain pits with a thin membrane across the opening and these presumably facilitate the diffusion of food materials.

Most of the genera possess systems of anastomosing mucilage ducts which are normally confined to the stipe, although in *L. saccharina* and *L. digitata* they enter the fronds as well. When mature there are periodic openings from these ducts to the exterior and their bases are lined with secretory cells. They arise lysigenously through an internal splitting of the thallus due to cell disintegration: this is followed by a differential growth so that the canals become more and more submerged in the thallus. The attachment organs or crampons, which are positively geotropic, have an apical growth and differ from the rest of the thallus in that there are no connecting hyphae nor is there any pith. The amount of conduction necessary in these plants would be expected to be small, but even so the degree of differentiation is remarkable. So far as the lamina is concerned the group is usually regarded as

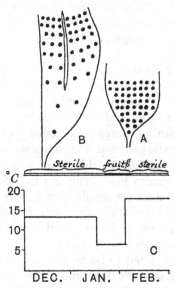

Fig. 118. *Laminaria.* A, *L. digitata,* marked thallus before growth in summer. B, *L. digitata,* marked thallus after growth in summer. C, effect of temperature on fruiting of gametophytes in *L. digitata.* (After Schreiber.)

primitive because the new portions do not originate separately but by intercalary growth from an existing portion (cf. fig. 115).

The sporangia and paraphyses are borne in irregular or more or less regular sori on both sides of the lamina. It is probable that the zoospores possess an eye-spot, but it must be very small because in the three species where it has been recorded it was very difficult to distinguish. The zoospores, which in one or two cases are reported to be of two sizes, germinate to form minute gameto-phytes, but on germination they first put out a tube that terminates in a bulbous enlargement into which the contents of the zoospore migrate. There the nucleus divides and one daughter nucleus passes into the tube whilst the other degenerates, but at present the significance of this phenomenon is obscure: it would hardly seem to be associated with meiosis because this process takes place in the zoosporangium. Both kinds of gametophyte show much variation in shape and size, the male gametophyte being the smaller throughout as it is built of smaller cells that contain dense chromatophores.

The gametophytes can be cultivated in the laboratory, but for successful cultivation the water must be sterilized and the cultures placed close to a north window in winter and 2 or 3 m. distant in summer. Reproductive organs are only formed at low temperatures, 2–6° C., whilst above 12–16° they are rarely produced, this fact perhaps accounting for their temperate and arctic distribution (cf. fig. 118). It is also known that the eggs may develop parthenogenetically to give a haploid sporophyte which has an irregular shape, whilst attempts to produce hybrids by artificial fertilization have so far met with no success. Schreiber (1930) found that the ratio of male to female gametophytes was always 1 : 1, and he subsequently showed that of the thirty-two zoospores produced in each sporangium sixteen gave male and the other sixteen female gametophytes. The male gametophyte of *L. religiosa* is reported to bear unilocular and plurilocular sporangia, but this is so abnormal and has never been confirmed or reported for any other species that it can hardly be accepted without further evidence. The ova of *L. saccharina* are reported to be capable of producing dwarf filamentous diploid plants which reproduce by means of unilocular sporangia. If this is confirmed it may be that here we have an example of a reversion to a primitive filamentous diploid progenitor, a feature which might help considerably in indicating their ancestry.

The most important characteristics of the gametophytic generation are:

(1) the male gametophyte always has smaller cells;

(2) the male gametophyte always consists of more than three cells whereas the female may consist of only one cell, the oogonium. Under good nutrient conditions both become much branched;

(3) the antheridia are unicellular and produce only one antherozoid;

(4) any cell of the female gametophyte may function as an oogonium;

(5) the male gametophyte degenerates after the gametes are shed whereas the female gametophyte persists.

The young sporophyte first produces numerous rhizoids of limited growth, but these are later covered by a disk-shaped expansion from which are produced the haptera or crampons.

LAMINARIACEAE: *Saccorhiza* (*sacco*, sack; *rhiza*, root). Fig. 119.

S. bulbosa used to be known as *Laminaria bulbosa*, but for some time it has been removed to a separate genus because it differs from the other species of *Laminaria* in several important respects.

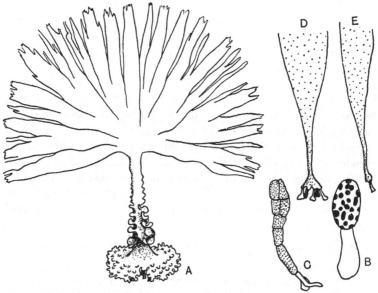

Fig. 119. *Saccorhiza bulbosa*. A, plant (× ⅓). B, female gametophyte. C, young sporophyte. D, E, young plants of *S. dermatodea* to show origin of bulb. (A, after Tilden; B, C, after Kniep; D, E, after Oltmanns.)

The persistent lamina arises from a flat compressed stipe with wavy edges which is twisted through 180° near the base as a result of unequal growth, this twisting being regarded as a mechanical device to facilitate swaying. The young sporophyte is attached at first by a small cushion-like disk, but later a warty expansion, the *rhizogen*, develops above it and forms a bulbous outgrowth which bends over and attaches itself to the substrate by means of descending crampons. As a result of the development of this adult holdfast the juvenile disk may be lifted completely off the substratum.

Subsequent growth of the stipe takes place in the outer layer of the medulla, and in the adult organ five regions can be recognized:

(1) Primary fixing organ.
(2) The bulb.
(3) A flattened twisted area said to provide additional rigidity.
(4) A portion with flounced edges.
(5) A flat straight portion that passes into the lamina.

The existence of these structures is supposed to be correlated with the large lamina which is cleft into many linear segments.

If, as sometimes happens, the whole of the plant is torn away with the exception of the bulb, this organ is still capable of reproduction and assimilation. The advanced external differentiation of the stipe is not reflected in its histology where the differentiation is poor because there is no secondary growing region, no mucilage ducts, and trumpet hyphae are not conspicuous.

Saccorhiza and *Alaria* are the only two genera in the Laminariales with cryptostomata that are at all comparable to those of the Fucales (cf. p. 194), the former genus possessing true cryptostomata with tufts of hairs. There are three theories concerning the homologies of the cryptostomata which may be mentioned briefly here (cf. also p. 196):

(1) They are incomplete sexual fucoid conceptacles which have failed to develop.
(2) They are forerunners of the sexual fucoid conceptacle.
(3) They are a parallel development with the sexual conceptacles of the fucoids, but otherwise have no relation to them.

Whilst there is very little evidence for any one of these theories it may be suggested that the second alternative probably fulfils most nearly the known facts.

The male gametophyte is filamentous whilst the female frequently consists of only one cell which functions as the oogonium. After fertilization has taken place the development of the sporophyte to maturity in both species requires only one year so that the plants are true annuals. *Saccorhiza bulbosa* is found on the Atlantic coasts of north and west Europe whereas the other species, *S. dermatodea*, is circumpolar and is possibly the parent species from

which the other developed, a speculation which is further supported
by the fact that *S. dermatodea* is more primitive because the stipe is
not twisted nor are the edges so wavy. The young sporophyte first
develops a juvenile blade which does not bear sporangia and then a
new and thicker basal fertile blade is intercalated, but it is only the
juvenile blade that bears the cryptostomata, thus suggesting that
these structures may be juvenile sexual conceptacles.

LAMINARIACEAE: *Thalassiophyllum* (*thalassio*, sea; *phyllum*, leaf).
    Fig. 120.

The perennial sporophyte is apparently composed of a spirally
twisted, fan-shaped lamina unrolling from a one-sided scroll

Fig. 120. *Thalassiophyllum clathrus*. A–F, developmental stages to show the
origin of the single scroll (× ⅔). G, adult plant. (After Setchell.)

without any mid-rib. A study of the embryonal stages, however, shows that the young plant is flat and bilaterally symmetrical. The two edges then curl up and the plant tears down the centre giving rise to two lateral scrolls each unrolling from a thickened outer margin, but as one of the scrolls soon ceases to develop the mature plant only possesses one scroll borne on a solid bifid stipe with the vestigial scroll on one of the branches. Slitting is represented by rows of small holes which commence to develop after the first tear has taken place.

LESSONIACEAE: *Lessonia* (after R. P. Lesson). Fig. 121.

The plants grow erect and form "forests" in relatively deep waters off the shores bounding the southern Pacific, reminding one

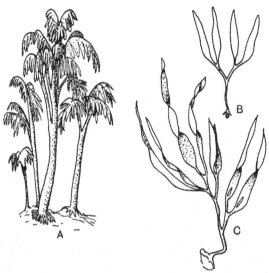

Fig. 121. *Lessonia.* A, adult plants of *L. fucescens.* B, C, seedling stages in *L. fucescens.* (After Oltmanns.)

in appearance of some of the fossil vegetation of the Carboniferous, although, of course, there is no connexion. The stipe is extremely stout and rigid, 5–10 ft. long and sometimes as thick as a human thigh. It appears to be more or less regularly branched in a dicho-tomous fashion, a feature which is brought about by the lamina being slit down successively to the intercalary growing region, each successive segment developing into a new lamina with its own portion of stipe. Dried parts of the stipe, which can easily be taken for pieces of driftwood, are used by natives to make knife handles. This method of causing splitting should be compared with the other processes found in *Nereocystis*, *Macrocystis* and *Postelsia* (cf. below).

LESSONIACEAE: *Postelsia* (after A. Postels). Fig. 122.

This is a monotypic genus, often known as the "sea palm", that is confined to the Pacific coast of North America where it grows between Vancouver Island and central California on rocks which

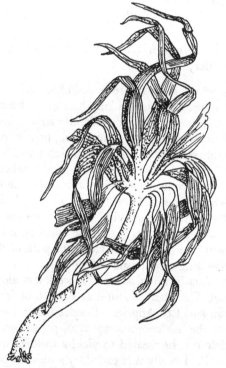

Fig. 122. *Postelsia palmaeformis*. (After Oltmanns.)

are exposed to heavy surf. The smooth, glossy, cylindrical stipe is thick but not very long, up to 1 m. in height. It is erect and hollow within and bears at its apex a number of short, solid, dichotomously branched structures from each of which hang 100–150 laminae that bear sporangia in longitudinal folds when they are mature. Apart from the cryptostomata of *Saccorhiza* and *Alaria* it has also been suggested that the occurrence of these sporangia in folds may illustrate how the fertile fucalean conceptacle may have arisen. Such a change would necessitate the development of wedges of sterile tissue in order to divide up the folds, but whether such a change could occur in a relatively differentiated thallus is a matter for speculation.

The numerous laminae are formed by a splitting process in which a portion of the lamina fails to continue growth whilst the rest goes on growing, and in this manner a weak area is formed from which a split commences.

LESSONIACEAE: *Nereocystis* (*Nereo, Nereis*, daughter of Nereus; *cystis*, bladder). Fig. 123.

The plants, which from the recorded observations appear to be annuals, may attain a maximum length of 90 m. bearing a bladder up to 2 or 3 m. in length. The long slender stipe is solid and cylindrical below but swollen and hollow above, finally contracting just below the terminal spherical bladder which bears a row of short dichotomous branches, each giving rise to a number of long thin laminae. The plant commences with only one blade which divides twice in a dichotomous fashion, thus producing four blades, and these form the centre of activity for the remainder through a process of slitting. The splitting of these four fronds is preceded by the development of a distinct line along the path of the future slit, the line representing new tissue, which has in consequence very little strength, thus forming an area of weakness along which the slit commences. The plant is found at a depth of from 5 to 25 m. between Alaska and Los Angeles. Besides being a good source of potash salts, as the ash contains 27–35 % potassium chloride, the stalk and vesicle can be treated to yield a candied edible product called "Seatron". Locally it is called by a number of names, bull kelp, bladder kelp, ribbon kelp and sea-otter's cabbage.

In the closely related genus *Pelagophycus* the spores are said to be non-motile, not even possessing cilia. Further confirmation of this fact is much to be desired because not only is it an unique state in the family but it also renders comparison with *Nematophyton* (cf.

p. 275) of great interest. Local names employed for *Pelagophycus* are elk-kelp, sea pumpkin and sea orange.

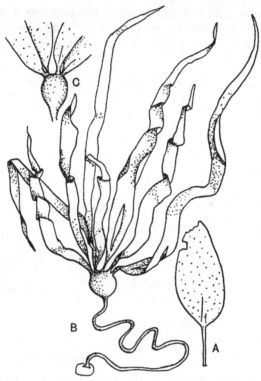

Fig. 123. *Nereocystis Luetkeana*. A, young plant. B, mature plant. C, branching from bladder. (After Oltmanns.)

*LESSONIACEAE: *Macrocystis* (*macro*, large; *cystis*, bladder). Fig. 124.

The perennial fronds of this giant of the ocean may reach 200 ft. in length, the alga growing at a depth of 20–30 m. in the North and South Pacific Ocean and near the Cape of Good Hope, all being regions where the temperature of the water ranges between 0 and 20° C. In the juvenile plant the stipe is simple and solid, but later on it branches one, two or three times in a dichotomous fashion, although ultimately the branching becomes unilateral and sympodial, each branch bearing two to eight laminae. The main growing region on each branch is ventrally situated in the terminal flag or blade, and it is here too that splitting takes place to form the individual laminae. The splitting is brought about by local gelatinization of the inner and middle cortex together with a cessation of growth in the epidermal area; this forces the adjacent tissues into the gelatinized areas until finally the epidermis is ruptured. Two

Fig. 124. *Macrocystis pyrifera.* A, young plant ($\times \frac{1}{4}$). B, slightly older plant with primary slit and two secondaries ($\times \frac{1}{4}$). C, still older plant. D, young plant. E, origin of blades at the apex. F, young plant. G, mature plant. H, sporangial sori ($\times \frac{1}{12}$). I, transverse section of thallus showing ridges ($\times 3.5$). J, surface view of holdfast of old plant showing flattened rhizome. (A, B, after Brandt; C, H, J, after Setchell and Gardner; D–G, after Oltmanns; I, after Smith and Whitting.)

kinds of zoospore are recorded, large ones which give rise to the female gametophytes and smaller ones which give rise to the male. The appearance of true heterospory in such an advanced alga is a feature of considerable importance because the phenomenon is normally associated with the land plants. The eggs are reported to be fertilized whilst still in the oogonium and if this is so then we have here the only example among the brown seaweeds of the retention of the ovum on the parent plant. This again may prove to be a significant feature in a consideration of the origin of a land flora.

ALARIACEAE: *Alaria* (*ala*, wing). Fig. 125.

This genus is widely distributed throughout the northern hemisphere, the common species being *A. esculenta*. There is a

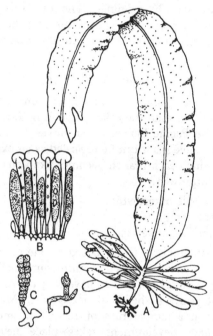

Fig. 125. *Alaria esculenta.* A, plant of *A. oblonga* with sporophylls. B, sporangia and paraphyses (× 200). C, germling sporophyte (× 100). D, female gametophyte (× 80). (A, after Oltmanns; B–D, after Newton.)

short, solid, unbranched stipe which is attached to the substrate by means of small branched rhizoids. It is naked below with an inter-calary growing zone that allows for continual renewal, whilst above the growing region the stipe expands into a flattened rachis which bears each year a fresh crop of marginal rows of sporophylls. The frond finally terminates in an expanded sterile lamina with a well-marked mid-rib, which is also an annual production. In addition to the intercalary growth there is also a marginal growth that imparts a wavy appearance to the terminal frond. This bears the so-called cryptostomata, although these are barely more than tufts of hairs arising in slight depressions. The sporangia are produced on the lower blades mixed up with unicellular paraphyses. The gametophytes are protonemal in form, simple or sparingly branched, the male, as usual, being composed of smaller cells with terminal, intercalary, or lateral antheridia, whilst the oogonia on the female gametophyte are usually terminal. The ovum is fertilized on emergence from the pore of the oogonium and the young sporophyte develops *in situ* without the characteristic early appearance of an holdfast.

ALARIACEAE: *Egregia* (outstanding). Fig. 126.

This genus is composed of two species, one having a more northern distribution than the other, though both are confined to the waters of the Pacific between Vancouver Island and Lower California. The whole plant can be regarded as an extension of the *Alaria* type in which each branch becomes strap-shaped and bears three types of outgrowth:

(*a*) Ligulate sterile outgrowths.
(*b*) Small fertile outgrowths.
(*c*) Conspicuous stipitate bladders.

The female gametophyte is composed of one or two large cells whilst the male plant is composed of numerous smaller ones, both plants reaching maturity in from 19 days to 4 weeks depending on the season of the year, e.g. the length of daylight. Maturity is most rapidly reached at a temperature of 10–16° C., and although at 16–20° C. gamete development takes place nevertheless the antherozoids are unable to leave the antheridia.

ALARIACEAE: *Eisenia* (after G. Eisen). Fig. 127.

The perennial sporophyte arises from a holdfast that is apparently bifurcate, although the two apparent branches are actually the lower margins of the primary lamina. The original elongate stipe, which may be as much as 15 cm. in length, is persistent and bears a flattened lamina from which pinnules develop. This primary lamina then disappears leaving two groups of pinnules or sporophylls attached to the lower and outer margin of the lamina side of the original transition area, whilst a small partial blade persists at the outer extremity of each false stipe. New sporangia continually arise at the base of the old ones, and the genus is interesting because the cuticle is shed when the sorus is mature (cf. p. 277). This is one of the few Laminariaceae

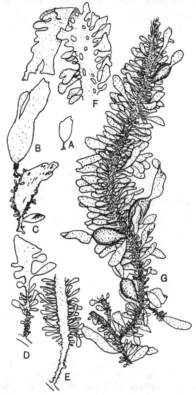

Fig. 126. *Egregia Menzesii*. A–C, stages in growth of young sporophytes (× $\frac{5}{12}$). D, young frond. E, base of mature frond. F, apex of mature frond. G, mature plant. (A–C, after Griggs; D–G, after Oltmanns.)

in which the number of chromosomes has been counted, the haploid number being fifteen. Two species are known, one from southern California and one from Japan.

ALARIACEAE: *Pterygophora* (*pterygo*, wing; *phora*, bearing).

The perennial sporophyte, which arises from a holdfast of branched haptera, possesses a simple, solid stipe that is more or less woody, being by far the stoutest known among the algae. The numerous linear laminae, about forty in number, are borne terminally, and though they have no distinct mid-rib nevertheless the central portion is much thickened. Long sporophylls are also produced laterally on both sides of the stipe near the transition area. These fronds, which possess continual growth, appear first in February and fruit in the following September or October, the

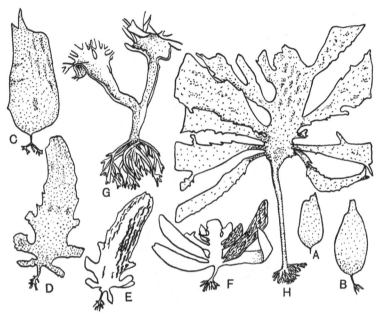

Fig. 127. *Eisenia*. A, young sporophyte of *E. bicyclis*. B–F, stages in development of the adult sporophyte of *E. arborea*. G, base of an adult plant of *E. bicyclis*. H, mature sporophyte of *E. bicyclis*. (After Tilden.)

sporangia and paraphyses being borne in sori on both sides of the sporophylls and also on the terminal laminae. *Pterygophora* is a monotypic genus found from Vancouver Island to Lower California where it grows characteristically at the bottom of deep chasms possessing 12–15 ft. of water at low tide. It has been estimated that individual plants may live for as long as 13 years.

REFERENCES

Desmarestia. ABE, K. (1938). *Sci. Rep. Tôhoku Univ.* IVth ser. **12**, 475.
Eisenia. HOLLENBERG, G. J. (1939). *Amer. J. Bot.* **26**, 34.
Laminaria. KANDA, T. (1936). *Sci. Pap. Inst. Alg. Res. Hokkaido Univ.* **1**, 221.
Laminaria. KYLIN, H. (1916). *Svensk bot. Tidskr.* **10**, 551.
Chorda. KYLIN, H. (1918). *Svensk bot. Tidskr.* **12**, 1.
Laminaria. SCHREIBER, E. (1930). *Planta,* **12**, 331.
Desmarestia. SCHREIBER, E. (1932). *Z. Bot.* **25**, 561.
Saccorhiza. SETCHELL, W. A. (1891). *Proc. Amer. Acad. Sci.* **26**, 177.
Eisenia. SETCHELL, W. A. (1896). *Erythrea,* **4**, 155.
Embryology, Regeneration. SETCHELL, W. A. (1905). *Univ. Cal. Publ. Bot.* **2**, 115.
General. WILLIAMS, J. LLOYD (1925). *Rep. Brit. Ass.* Pres. Address, Sect. K, p. 182.

CYCLOSPOREAE—
FUCALES (HETEROGENERATAE)

The sporophytic plants are even more dominant in the life cycle than in the Laminariales, but although diploid there is no apparent asexual reproduction, the plants always reproducing by means of ova and antherozoids. There is considerable tissue differentiation, and in their external features the plants exhibit much more variation than is to be found in the Laminariales. Some workers consider that the structures called oogonia and antheridia are really macro- and microsporangia producing mega- and microspores which germinate before they are liberated from the sporangium, so that while the reproductive bodies have their origin as spores, nevertheless the liberated products are gametes. This view is held by the present author and is discussed more fully later (cf. p. 258). In the primitive condition eight ova are produced in each oogonium and sixty-four antherozoids in each antheridium. Meiosis takes place during the first two divisions in the formation of microspores, and as there is often a pause after the second division the first four nuclei have been regarded as the functional microspores, each of which subsequently undergoes four mitoses so that they can be said to germinate to a sixteen-celled gametophyte where each cell functions as an antherozoid. In the macrosporangium the first four nuclei formed are regarded as the functional megaspores, and each of these is considered to germinate subsequently to a two-celled female gametophyte where each cell functions as an ovum. In those species where less than eight mature ova are produced it must be assumed that some of the megaspores undergo abortion.

If the above is to be the correct interpretation, and it would seem to be more satisfactory than any other theory in comparison with other members of the Phaeophyceae, then we can say that not only is there a cytological alternation of generations but there is also a morphological alternation, although the sexual generation is even further reduced from the state found in the Laminariales. This really forms the basis for placing the Fucales in the Heterogeneratae. The alternative interpretation is that the sexual generation has been completely suppressed and is solely represented by the gametes, so that whilst there is a cytological alternation of generations there is only one morphological generation (cf. also Chapter IX). The sporangia are borne in flask-shaped depressions of the thallus called conceptacles, each of which is lined with paraphyses and opens to the surface by means of an ostiole. The plants of the different species may be dioecious, monoecious or hermaphrodite. It has been pointed out that the number of primary rhizoids in the embryo is proportional to the size of the rhizoidal cell, which in turn bears a relation first to the size of the egg, and secondly to the complexity of the thallus. On this basis a series of increasing embryonal complexity may be traced, e.g. *Fucus → Ascophyllum → Pelvetia → Cystoseira → Sargassum*.

Geographically the original centre of distribution was undoubtedly the southern Pacific in the waters of Australia and New Zealand where the greatest number of species are now to be found. This makes an interesting comparison with the preceding order whose original centre of distribution was the northern Pacific in the waters around Japan and Alaska. The Fucales are classified into five groups, the classification being based primarily upon the structure of the apical growing cell or cells:

(1) *Durvilleaceae.* A group comprising two genera, *Durvillea* and *Sarcophycus*, from Australia and Patagonia, both without any means of apical growth.

(2) *Fuco-Ascophyllae.* Growth is determined in the adult stage by *one four-sided* apical cell.

(3) *Loriformes.* Growth is due to *one three-sided* apical cell which gives rise to a long whip-like thallus.

(4) *Cystoseiro-Sargassae.* The apical cell is again *three-sided* but

there is copious branching which results in bilateral, radial and bilaterally radial thalli.

(5) *Anomalae*, composed of two genera, *Hormosira* and *Notheia*, both confined to the Antipodes. Growth is brought about by a group of cells instead of a single cell.

DURVILLEACEAE

Durvillea (after I. D. D'Urville). Fig. 128.

The sporophyte is a dark olive brown or black in colour and possesses very much the appearance of a *Laminaria*. The large

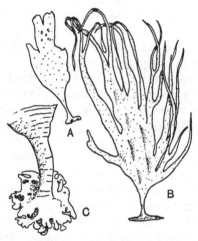

Fig. 128. *Durvillea antarctica*. A, young plant (× ⅓). B, adult plant (much reduced). C, stipe and holdfast (× ½). (After Herriot.)

solid stipe arises from a scutate holdfast and very soon passes into a flat, expanded, fan-shaped lamina, which later becomes split into segments although no definite appendages are produced from this frond. The ends of the older laminae become frayed and broken off by wave action, whilst the holdfast may attain a diameter of 2 ft. through the addition of new tissue annually. If this secondary growth did not occur the plant would soon be torn from its moorings because the holdfast is continually becoming riddled with holes through the boring operations of molluscs. The macro- and microsporangia, which are borne in conceptacles on different plants as the genus is dioecious, occur over the whole of the lamina, this condition being regarded as the primitive state for the Fucales. It is known as the "bull kelp" and forms submarine forests in deep waters off New Zealand and the Aucklands down to depths of 30 ft., or else it grows in places continually exposed to surf.

FUCO-ASCOPHYLLAE

*FUCACEAE: *Fucus* (a seaweed). Figs. 129–131.

This genus contains a number of species that are widely scattered over the world with the majority in the northern hemisphere, many of them exhibiting a wide range of form with numerous so-called varieties. When two or more species occur in the same area they are generally present in different zones on the shore, probably dependent upon the degree of desiccation that they can tolerate (cf. p. 353). The plants are attached by means of a basal disk and there is usually a short stalk, which continues on to form the mid-rib of the frond in those regions where the expanded wings or *alae* are developed, these latter being of varying width with either entire or serrate margins. Branching is commonly dichotomous or subpinnate, and in many species the branches bear expanded vesicles or *pneumatocysts*. Sometimes whole portions of the frond may be inflated in an irregular manner, but the factors causing this phenomenon are not known, although it is possible that contact with rock or soil provides the necessary stimulus. With increasing age the lower portions of the alae may be frayed off by wave action, leaving only the mid-rib, which then has the appearance of a stipe. The whole of the expanded thallus is covered with sterile pits or crypto-stomata similar to those of *Saccorhiza*, but in fruiting plants it is only the ends of the branches that become swollen and studded with the fertile conceptacles. In *F. spiralis* these conceptacles are hermaphrodite, containing both mega- and microsporangia; in *F. vesiculosus* and *F. serratus* the plants are dioecious, the two types of sporangia occurring on separate plants, whilst in *F. ceranoides* either state may be found. A number of very peculiar forms have been described which commonly occur on salt marshes: these rarely fruit, reproduction being secured principally by means of vegetative proliferations (cf. p. 325). The age of *Fucus* plants has not been studied in much detail but the following figures (Table I) may be cited from one worker who marked a number of plants:

TABLE I

Species ...	F. spiralis	F. serratus	F. vesiculosus	Ascophyllum nodosum
Max. age (yr.)	3½	4	2½	2½
Av. age (yr.)	1½	2	1	1½

Morphologically the thallus shows considerable differentiation. The external layer, which is known as the *limiting layer*, consists of small cells with abundant plastids and is primarily assimilatory in function. Below this there is a *cortex* composed of several layers of parenchymatous cells which become more and more elongate and mucilaginous towards the centre, and these probably form the storage system. In the very centre the cells are extended into hyphae which are interwoven into a loose tangled web. This central tissue is called the *medulla* and probably acts as a conducting system, because the transverse walls of the hyphae are frequently perforated with the same type of pit that is to be found in some of the Laminariaceae. The primary medullary hyphae are relatively thin-walled, but when secondary growth of the thallus takes place the new hyphae which result from this process are very thick-walled and so are probably mainly mechanical in function. Secondary growth is due to the activity of the limiting layer and the inner cells of the cortex, the latter tissue being responsible for the formation of the secondary hyphae (cf. fig. 131) which penetrate between the primary medullary hyphae and finally outnumber them. There is a greater development of secondary thickening in the stipe and mid-rib than there is in the frond, whilst in very old parts of the thallus the limiting layer may die off and then the underlying cortical cells take over its function.

Growth in length takes place by means of an apical cell which lies at the bottom of a slit-like depression that has resulted from the more rapid growth of the surrounding limiting layer. The apical cell is three-sided in young plants whilst in the adult thallus it becomes four-sided, the new segments being cut off successively from the base and four sides, after which they develop into the various tissues (fig. 129). Injury, and also the stimulus provided when the thallus lies on marsh soil, induces new growth in the neighbouring cells, and in this manner proliferations are formed which may also serve for vegetative propagation. Both crypto-stomata and conceptacles arise as depressions in the surface of the

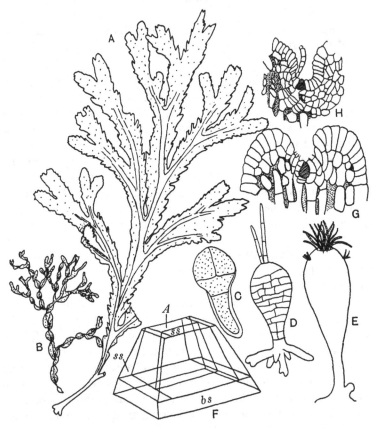

Fig. 129. *Fucus*. A, adult plant of *F. serratus* (×0·30). B, a marsh form of *F. vesiculosus* (×0·30). C–E, seedling stages of *F. vesiculosus* showing origin of rhizoids and apical tuft of hairs. F, diagram to show method of segmentation of apical cell, *A*. *bs*=basal segment, *ss*=side segments. G, apical cell of young thallus. H, apical cell of old thallus. (A, B, after Taylor; C–H, after Oltmanns.)

thallus and there are three principal accounts which have been given of the course of their development:

(1) An early view held by Kützing and Sachs in which they were described as arising as slight depressions in the thallus that later

became overgrown by the surrounding tissue. This has since been abandoned.

(2) According to the second account a *linear series* of two or more cells is formed but their horizontal activity then ceases, thus leaving a terminal initial cell which becomes sunk in a depression as the surrounding tissues grow up. On this theory the sides of the conceptacle are derived from the limiting layer and underlying cortex, as Bower (1880) demonstrated for *Fucus*, whilst in *Himanthalia* the

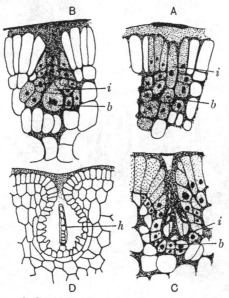

Fig. 130. *Fucus.* A–C, origin of conceptacles in *F. serratus.* *b* = basal cell, *i* = initial. D, juvenile conceptacle of *Cystoseira.* *h* = hair. (After Oltmanns.)

sides are derived from the limiting layer only. Finally, around the remnants of the one or more initial cells a central mucilaginous column is formed stretching to the neck of the conceptacle and connected to the walls by thin strings of mucilage which are later ruptured. According to this description, therefore, the conceptacles are the products of one or more initials which may or may not disintegrate at a later stage (cf. fig. 130).

(3) The third account describes the conceptacle as developed entirely from a *single initial* that divides transversely into two unequal cells, the upper or tongue cell degenerating whilst the lower

one gives rise to the walls of the conceptacle. This method of formation has been successfully demonstrated for *Sargassum*, *Pycno-
phycus* and other Fucaceae. It is clear from the investigations that
have been made that both methods (2) and (3) are to be found in the
different species.

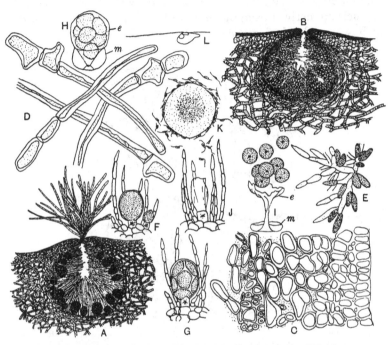

Fig. 131. *Fucus.* A, transverse section "female" conceptacle of *F. platycarpus*.
B, transverse section "male" conceptacle of *F. vesiculosus*. C, portion of thallus of
F. spiralis to show structure (× 125). D, origin of hyphae at 1 cm. below apex in
F. spiralis (× 235). E, microsporangia. F, young, and G, old megasporangium.
H, liberated ova in inner vesicle. *e* = endochiton, *m* = mesochiton. I, ova being
liberated. *e* = endochiton, *m* = mesochiton. J, empty sporangium showing torn
exochiton. K, ovum being fertilized. L, antherozoid. (C, D, after Pennington;
rest after Oltmanns.)

The cryptostomata or hair pits are regarded as a juvenile stage of
the fertile conceptacle (cf. also p. 178) because sporangia are
frequently associated with the hairs or else they occur in the same
cavity after the hairs have been lost. With this interpretation in
view the following morphological series can be arranged:

(a) Plants with a continuous patch of hairs and reproductive bodies, e.g. *Laminaria*.

(b) Plants with hairs and reproductive bodies in scattered sori, e.g. *Dictyota*.

(c) Plants with hairs and reproductive bodies in scattered receptacles, e.g. *Durvillea*.

(d) Plants with hairs and reproductive bodies in receptacles which are confined to apical positions or special side branches, e.g. *Fucus, Ascophyllum*.

In the mature fruiting conceptacles there are branched hairs or paraphyses with the microsporangia borne terminally on the branches near the base, or else the paraphyses are unbranched and associated with the megasporangia, which are either sessile or else borne on a single stalk cell, each megasporangium characteristically containing eight ova when mature. In those species where the conceptacles are hermaphrodite all these structures occur together. The walls of both sporangia are double, and when the gametes are ripe the sporangia burst, liberating their contents which are still enclosed in the inner delicate membrane. The expulsion of the gametes normally takes place whilst the tide is out because the conceptacle is then full of mucilage and the loss of water causes the thallus to shrink, thus forcing the ripe ova and antherozoids in their envelopes through the ostiole to the surface. When the tide returns the inner wall bursts and so liberates the antherozoids, whilst the inner megasporangium wall inverts and enables the ova to escape. Fertilization takes place in the sea, the antherozoids clustering around the ova and causing them to rotate by their activity until one antherozoid succeeds in entering and fertilizing each ovum.

The fertilized zygote surrounds itself with a wall and very shortly begins to divide, the direction of the first wall being said to be at right angles to the incident light. After a few more divisions the octant stage is reached and then a rhizoid appears on the side away from the light and grows downward, being followed soon after by others (cf. p. 289). The upper part of the embryo elongates from a five-sided apical cell but the end soon becomes flattened, after which a terminal depression arises that contains the three-sided juvenile apical cell together with a bunch of hairs. The bunch of hairs possess trichothallic growth, but they soon fall off and the basal cell of one hair becomes the new four-sided apical cell of the adult plant. It is perhaps of interest to note that in *Fucus vesiculosus* it has been shown that the mature sporophyte contains the diploid number of sixty-four chromosomes, which appears to be the usual number in all the Fucales so far examined, with the exception of *Sargassum Horneri* in which $2n = 32$.

*FUCACEAE: *Pelvetia* (after the French botanist, Dr Pelvet). Fig. 132.

The fronds in this genus have no mid-rib and are linear, compressed or cylindrical with irregular dichotomous branching. Air

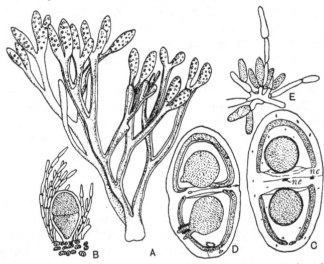

Fig. 132. *Pelvetia canaliculata*. A, plant ($\times \frac{3}{8}$). B, megasporangium ($\times 60$). C, mature fertilized sporangium ($\times 72$). *ne* = rejected nuclei. D, germinating oospores ($\times 72$). E, microsporangia ($\times 156$). (A, original; B–E, after Scott.)

vesicles may be present in some species but normally they are absent, especially in the European *P. canaliculata*, which grows on rocky shores forming a zone near high-water mark or even above so long as it is reached by the spray. Modified salt-marsh forms derived from *P. canaliculata* are also recorded but these are confined to Great Britain (cf. p. 324); like the marsh forms of *Fucus* they are characterized by the general absence of fruiting receptacles, reproduction being primarily vegetative. The structure of the thallus is essentially similar to that of *Fucus*, but the Californian *Pelvetia fastigiata* also possesses a few cryptostomata which are otherwise absent from the genus. The sporangia are similar to those of *Fucus* except that normally only two ova mature, the remaining six nuclei being extruded from the cytoplasm into the wall, though in *Pelvetia fastigiata* one may occasionally find four ripe ova or else ova that contain two nuclei. In *P. canaliculata* the two mature eggs are arranged one above the other, whilst in the Japanese species, *P. Wrightii*, they are placed side by side. This difference is probably dependent upon the relative position of the two megaspores which germinate.

*FUCACEAE: *Ascophyllum* (*asco*, wine-skin; *phyllum*, leaf). Fig. 133.

The plants of this genus are large, often attaining several feet in length, and are commonly to be found on sheltered coasts at about

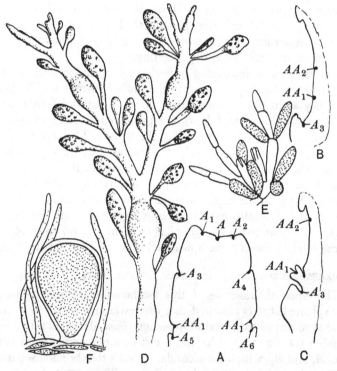

Fig. 133. *Ascophyllum nodosum*. A–C, diagram showing method of branching. *A*, apical cell. A_1–A_6, secondary initials in order of development. AA_1, AA_2, tertiary and quaternary initials. D, plant ($\times \frac{1}{2}$). E, microsporangia ($\times 225$). F, megasporangium ($\times 2\cdot 25$). (A–C, after Oltmanns; D–F, after Newton.)

mean sea-level. The thallus of the common species, *A. nodosum*, which sometimes bears nodular galls caused by the eel-worm *Tylenchus fucicola*, is more or less perennial, and regenerates each year from a persistent base or from the denuded branches. As in the two previous genera free-living or embedded forms have evolved in salt-marsh areas (cf. p. 324), and these differ considerably from the common parent species, *Ascophyllum nodosum*, not only vegetatively but also in the absence of sporangia. The normal fronds have a serrated margin but no mid-rib and commonly bear vesicles which are known as *pneumatocysts*, but when the vesicles are borne on the little side branches they are termed *pneumatophores*. The axis is beset by simple, clavate, compressed branchlets that arise singly or in groups in the axils of the serrations. These are later converted into or are replaced by short-stalked, yellow, fertile branches which fall off after the gametes have been liberated from their conceptacles. The macrosporangia each give rise to four ova, the remaining four nuclei degenerating.

The method of branching is perhaps best understood from an inspection of fig. 133. In spring the main branches divide dichotomously as in *Fucus*, after which opposite pairs of fertile receptacles or sterile tufts of hairs are produced in notches that are formed as follows on both sides of the thallus. The apical cell (A) cuts off another apical cell (A_1) that remains dormant for a time, during which period it is carried up the edge of the groove to the side of the thallus by the activity of the primary apical cell. The limiting layer immediately around A_1 does not undergo further growth and so it also comes to lie in a groove. Later on, tertiary (AA_1) and quaternary (AA_2) apical cells are cut off from A_1, the tertiary cell becoming the apical cell of a sterile or fertile branch.

FUCACEAE: *Seirococcus* (*seiro*, chain; *coccus*, berry). Fig. 134.

The mode of branching in this southern-hemisphere genus can be explained if it is assumed that the lower side of a notch, comparable to one of those found in *Ascophyllum*, develops into a leafy member (cf. fig. 134). The apical cell cuts off segments on either side, A_1 and A_2, which are secondary apicals that become separated from A through growth of the epidermis. These secondary apicals divide to give tertiaries, A_3, after which they become separated from each other by a new leaf organ (l) that develops as a result of the activity of one of the tertiary apicals. Subsequently the secondary apical, A_1, undergoes a series of divisions, thus producing a row of apical cells each of which develops into a fertile branchlet. The tertiary apical normally only gives rise to the leaf blade, but it may divide again sometimes to give a new shoot or a series of fertile

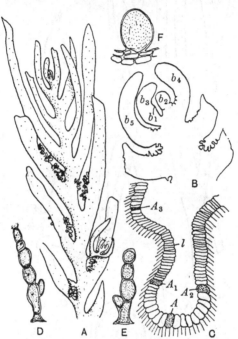

Fig. 134. *Seirococcus.* A, plant of *S. axillaris* with fruiting laterals. B, diagram to show method of branching. b_1-b_5 blades, b_1 being the youngest. C, diagram showing disposition of apical meristematic cells, $A-A_3$, the former being the oldest: l = origin of leaf organ. D, E, paraphyses of female conceptacle ($\times 135$). F, megasporangium ($\times 135$). (A–C, after Oltmanns; D–F, after Murray.)

branchlets which will thus appear to grow out from the main thallus.

LORIFORMES

FUCACEAE: Himanthalia (himant, thong; *halia,* of the sea). Fig. 135.

The short, perennial frond or button arises from a small disk-like holdfast, the shape of the button being dependent upon level because it is short and stumpy when it grows exposed at high

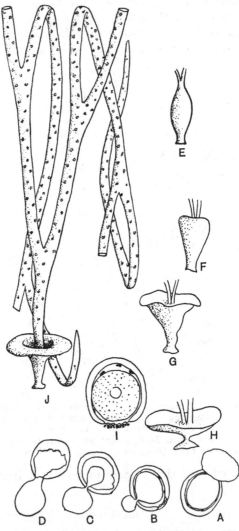

Fig. 135. *Himanthalia lorea.* A–D, stages in the liberation of the ovum (×22). E, F, abnormal buttons. G, button from bottom of dense zone. H, button from top of dense zone. I, mature megasporangium. J, plant with fertile fronds. (A–I, after Gibb; J, after Oltmanns.)

levels, whilst it is more elongate at the lower levels where the plants are submerged for longer periods. From March to July of each year new receptacles grow out from the centre of the buttons and form very long strap-shaped and repeatedly forked structures filled with mucus. Growth curves show that the greatest length is attained by these annual fronds on plants growing in the lowest part of the dense zone and that the shortest occur in the highest. This can be correlated with (a) the greater degree of desiccation at the higher levels, and (b) the fact that the less frequent flooding reduces the supply of available salts. Reduction has proceeded so far in this genus that only one ovum matures in the ripe macro-sporangium. The liberation of the gametes is controlled by the tides and exposure and there is a definite periodicity related to these two factors.

CYSTOSEIRO-SARGASSAE

SARGASSACEAE: *Halidrys* (*hali*, sea; *drys*, oak). Fig. 136.

The perennial fronds arise from a conical holdfast and bear pedicelled air vesicles, but as these are lanceolate and jointed they probably represent a series of vesicles. There are only two species, the European *H. siliquosa* being hermaphrodite whilst the Californian one is dioecious. In both, the stalked receptacles form terminal racemes at the apices of branches, but only one ovum matures in each macrosporangium. In the Californian *H. dioica* there are a number of interesting morphological features:

(a) An unbroken series can often be found which shows every gradation between a leafy member and the series of vesicles.

(b) Protoplasmic connexions between cells are continuous throughout the whole of the plant, a feature which should be compared with the condition commonly found in the Rhodo-phyceae (cf. p. 212).

(c) The origin of the vesicles appears to be largely dependent upon the food supply.

(d) The cells in the centre of the mid-rib have definite sieve plates comparable to those in the trumpet hyphae of *Laminaria*, though without the bulbous swellings.

(e) The air chambers and primary hyphae appear to arise in regions which are losing their vitality, though the significance of this behaviour is not clear.

Fig. 136. *Halidrys siliquosa*. A, plant (× ⅔). B, apex to show branching.
a = primary initial, *a₁*–*a₄* = secondary initials. (A, original; B, after Oltmanns.)

SARGASSACEAE: *Cystoseira* (*cysto*, bladder; *seira*, chain). Fig. 137.

The much-branched perennial thallus is either cylindrical or compressed and arises from a fibrous woody holdfast which has more or less the structure of a conical cavern. The primary branches arise from the main stipe towards the base and divide above into filiform branches and branchlets, but when the latter do not develop very far one gets what is known as the "*Erica*" and "*Lycopodium*" types, so called because of their resemblance to members of those genera. Seriate rows of small air vesicles may be inserted in the branches, and when this occurs the row of vesicles must be regarded as a modified branch. The plants are monoecious or dioecious, the conceptacles being borne in terminal or intercalary positions on the ramuli, and, as in some of the other genera, only one ovum develops in each megasporangium, the remaining seven nuclei degenerating. In the seedling the main shoot is very short and soon stops growth, and as a consequence it is completely overtopped by

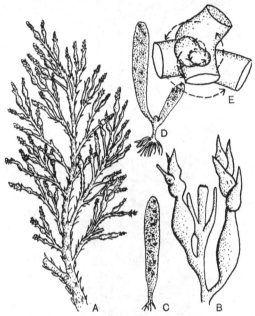

Fig. 137. *Cystoseira*. A, *C. ericoides*, plant (×½). B, portion of same enlarged (×4·5). C, germling. D, same, rather older. E, diagram to show nature of branching in *C. abrotanifolia*. (A, B, after Newton; C–E, after Oltmanns.)

the lateral branches. The first two shoots arise opposite each other but the remainder have a divergence of 2/5. The genus is principally confined to the warmer subtropical and temperate waters of the globe.

*Sargassaceae: *Sargassum* (*sargasso*, Spanish for sea-weed). Fig. 138.

The branching in this genus is radial with a divergence of 2/5. The primary branch is a sterile phylloclade which bears crypto-stomata whilst the secondary branch is also sterile and is commonly reduced to an air-bladder. In some species there may be yet a third sterile branch which is also reduced to an air-bladder, but all the subsequent branches are fertile and finger-like in appearance. The plants are attached by means of a more or less irregular, warty,

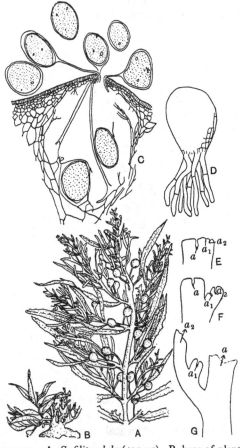

Fig. 138. *Sargassum*. A, *S. filipendula* (× 0·45). B, base of plant. C, escape of sporangia each with eight nuclei (× 40). D, seedling at rhizoid stage (× 105). E–G, stages in branching, *S. Thunbergii* (× 0·22). a = main initial, a_1 = branch initial, a_2 = secondary branch initial. (A, B, after Taylor; C, after Kunieda; D, after Tahara; E–G, after Oltmanns.)

solid, parenchymatous base or else numerous stolon-like structures grow out from the main axis and anchor the plant. The genus, which is principally confined to tropical waters, is a very large one with about 150 species, some being dioecious whilst others are monoecious. In the ripe megasporangium only one ovum reaches maturity under normal conditions, though occasionally eight eggs may develop. In the former case the single ovum contains all eight nuclei, but only one of these grows larger and is actually fertilized. This state of affairs can be interpreted as a failure on the part of the megaspores and gametophytes to form cell walls, and is a secondary condition due to still further reduction. In *S. filipendula* there is no stalk to the megasporangium and so it is embedded in the wall of the conceptacle. When ripe the whole megasporangium, not merely the inner wall and its contents, is discharged and remains just outside the ostiole attached to the conceptacle wall by a long mucilaginous stalk.

After fertilization the first divisions take place whilst the zygote is still attached to the parent plant by this long stalk. In *S. filipendula* fertile sporangia or degenerate sporangia are found in some of the cryptostomata, and this fact has been taken to signify that these sterile pits are abortive or juvenile conceptacles. The genus is especially abundant in Australian waters, one species, *S. enerve*, being employed in Japan as a decoration for New Year's Day because, when dried, it turns green. Various species are also used in the same country for food, but the chief claim to notoriety in this genus is probably associated with *S. natans*, the so-called Sargasso weed, which from time immemorial has been found as large floating masses in the Sargasso Sea near the West Indies, frequent references to it being recorded in the stories of early travellers to that region. At one time it was thought that plants of *S. natans*, together with one or two other species that behave similarly, were attached in the early stages, but there would now seem to be good evidence that they remain floating throughout the whole of their life cycle. Börgesen suggests that these perennial pelagic species originally arose from attached forms such as *S. vulgare*, *S. filipendula* and *S. Hystrix*.

SARGASSACEAE: *Turbinaria* (like a spinning top). Fig. 139.

The dioecious sporophyte forms a cone-like bush up to 25 cm. high arising from a branched holdfast. The stiff cylindrical stipe is crowded with leaves which are triangular or disk-like structures borne on petioles that represent the primary sterile branch of

Sargassum, these leafy bodies serving not only as assimilatory organs but also as floats. All the subsequent branches, which grow in corymbose clusters from the base of the phylloclade, are fertile. The genus is essentially confined to the warm waters of the tropics and subtropics.

Fig. 139. *Turbinaria.* Portion of plant with sterile (*s*) and fertile (*f*) branches. (After Oltmanns.)

ANOMALAE

*FUCACEAE: *Hormosira* (*hormo*, necklace; *sira*, a chain). Fig. 140.

The sporophyte, which has the appearance of a bead necklace, is composed of a chain of swollen vesicles (internodes) connected by narrow bridges (nodes). Growth takes place by means of a group of four apical cells, and these give off branches alternately in a dichotomous manner, the branches usually arising at the internodes; but apart from the discoid holdfast, there is no differentiation into appendages comparable to those of the other Fucaceae. The basal internode is solid but all the remainder are hollow: the nodes are also solid because they are composed solely of epidermis and cortex. The sporophytes are dioecious, the conceptacles being borne on the periphery of the inflated nodes. Although eight ova are originally formed in the megasporangia only four attain to maturity, but in this genus, however, it is a case of degeneration of eggs and not merely of nuclei. Another interesting feature of this genus is its capacity to form and shed a cuticle that bears the im-

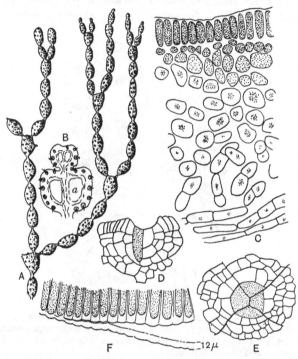

Fig. 140. *Hormosira Banksii.* A, portion of plant ($\times \frac{2}{5}$). B, longitudinal section of apex of plant. a = air-filled space. C, transverse section of thallus at internode ($\times 150$). D, longitudinal section of apex. E, transverse section of apex. F, cuticle being shed (semi-diagrammatic). (A, C, after Getman; B, D, E, after Oltmanns; F, original.)

pressions of the cell outlines, this feature perhaps being of signi-ficance when the problems concerning the Nematophyceae are considered (cf. p. 277). The genus is monotypic, the single species, *H. Banksii*, being confined to Australia and New Zealand where it grows on rocks and in tide pools of the littoral belt in positions that are always exposed to the spray.

FUCACEAE: *Notheia* (a spurious thing). Fig. 141.

The filiform sporophyte grows out parasitically from the base of conceptacles in *Hormosira* and *Xiphophora*, though it is present most commonly on the microsporangiate hosts. The thallus is solid throughout and is composed of epidermal, cortical and medullary tissues, the epidermis, like that of *Hormosira*, possessing a cuticle. The genus differs from *Hormosira* in that growth is secured by a

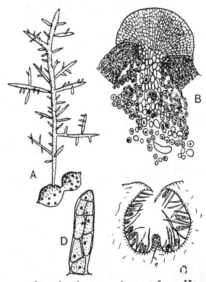

Fig. 141. *Notheia anomala*. A, plant growing out from *Hormosira*. B, point of entrance of parasite into host (× 40). C, conceptacle with megasporangia and branch shoot, *s* (× 40). D, mature megasporangia with eight ova (× 180). (A, after Oltmanns; B–D, after Williams.)

group of three apical cells instead of four. Branching is irregular, the new branches arising in the walls of old conceptacles from basal cells which were dormant during the reproductive phase. There is a degenerate holdfast which is composed of colourless elongated cells that penetrate the host and act as absorbing organs, although there are no actual haustoria. In those portions of *Hormosira* that are attacked by the parasite the hollow of the vesicle-like internode becomes filled up by new tissue formed as a result of the stimulation, but the parasite is apparently unable to attack *Hormosira* unless the host is growing in areas where it is continuously submerged. The fertile conceptacles, which only contain megasporangia, cover the entire frond, but as microsporangia have never been recorded the occurrence of meiosis is extremely doubtful, although at present there is no cytological evidence available. Each mature megasporangium, which contains eight eggs, is surrounded by unbranched paraphyses. The genus is monotypic and contains the one species, *Notheia anomala*, which, in view of its habit, structure and life history, must be regarded as a degenerate type.

REFERENCES

Sargassum. BÖRGESEN, F. (1914). *Mindeskr. Steenstr.* p. 3.
Fucus. BOWER, F. O. (1880). *Quart. J. Micr. Sci.* **20**, 36.
Hormosira. GETMAN, M. R. (1914). *Bot. Gaz.* **58**, 264.
Himanthalia. GIBB, D. C. (1937). *J. Linn. Soc. (Bot.)* **51**, 11.
Fucus, Pelvetia. INOH, S. (1935). *J. Fac. Sci. Hokkaido Univ.* **4**, 9.
Fucus. NIENBURG, W. (1931). *Wiss. Meeresuntersuch.*, Abt. Kiel, **21**, 51.
Fucus. ROE, M. L. (1916). *Bot. Gaz.* **61**, 231.
Sargassum. SIMONS, E. B. (1906). *Bot. Gaz.* **41**, 161.
Notheia. WILLIAMS, M. M. (1923). *Proc. Linn. Soc. N.S.W.* **48**, 634.

CHAPTER VIII

RHODOPHYCEAE

*Systematically these form a large but very uniform group in so far as their reproductive processes are concerned, although they may vary widely in the construction of the vegetative thallus. As in the Chlorophyceae there is one section that is characterized by lime encrustation, these algae having played a great part during past geological ages in the building up of rocks and coral reefs (cf. p. 273), a process which can still be seen going on in the tropical seas to-day. Morphologically the thallus is built up on one of two plans:

(*a*) Central filament type in which there is a central corticated or uncorticated main axis bearing the branches (fig. 144, B).

(*b*) Fountain type in which there is a mass of central threads all of which lead out like a spray to the surface, e.g. *Corallina*.

The cells composing the plants are frequently multi-nucleate, and contain, in addition to the components of chlorophyll, the red pigment phycoerythrin together with phycocyanin in some cases, whilst *Polysiphonia* is interesting in that it also contains fucoxanthin. With the exception of the first subdivision, the Protoflorideae, the cells remain united to one another after segmentation by means of thin protoplasmic threads or *plasmodesmae*, which are very conspicuous in the region of the fusion cell (cf. below), where their size can be associated with the need for the transmission of nutritive material. The reproductive bodies are very characteristic, usually being found on separate plants, but the two sex organs may occur on the same plant and certain abnormal cases are also known where sexual and asexual organs are present on the same thallus (cf. p. 236). The sexual plants are usually all of the same size, but in *Martensia fragilis* and *Caloglossa Leprieurii* the male plants are smaller than the female.

The male organs, which are probably best termed antheridia although they have been given other names, each give rise to a non-motile body, or *spermatium*, which is carried by the water to the elongated tip (*trichogyne*) of the *carpogonium* or female organ. In this respect it will be seen that the Rhodophyceae are very distinct from

RHODOPHYCEAE 213

the other algal groups. The carpogonium with its trichogyne is borne on a special branch (*procarp*) consisting of a varying number of cells, whilst the typical *auxiliary cell*, into which the fertilized carpogonial nucleus generally passes, is often associated with this branch, or else forms a part of it. The fertilized zygote commonly gives rise first to a peculiar diploid generation, the *carposporophyte*, which consists of a series of filaments that cut off asexual bodies or *carpospores* from their apices. These spores on germination usually give rise to the asexual or *tetrasporic* plant which reproduces by means of tetraspores that are formed in sporangia borne externally or else sunk into the thallus (cf. fig. 155, F). A common feature in this group that further emphasizes their uniformity is a tendency for the 2*x* number of chromosomes to be 40. The Rhodophyceae may be regarded as the classical example of plants in which meiosis occurs at different phases in the life cycle, for it may either occur immediately after fertilization or else at some subsequent period. In the former case the plants are said to be *haplobionts* as there is only one kind of individual or biont, but the *individual* sexual haploid plants are termed *haplonts*. If meiosis is delayed we get an asexual generation alternating with the sexual and so there are two kinds of individuals or bionts: this type is therefore known as *diplobiontic*. It may be pointed out here that other usages of these terms have been employed, but the above definitions are those propounded by Svedelius (1931) who coined the terms, and therefore they are the correct way in which they really should be employed. The classification of the Rhodophyceae is based primarily on the structure of the female reproductive apparatus. After the Proto-florideae, which lack pit connexions, have been segregated, the remainder of the red algae, or Eu-florideae, are classified as follows:

(1) *Nemalionales* and *Gelidiales*.

These are regarded as primitive orders which have become more or less stabilized. In some genera there are no true auxiliary cells, whilst in others the auxiliary cells are purely nutritive, but nevertheless the beginnings of an evolutionary series can be seen in the following features:

(*a*) The development of the hypogenous cells of the carpogonial branch to form storage organs.

(*b*) The development of special nutritive cells which will ultimately replace the auxiliary cells in fulfilling the nutritional requirements.

(*c*) The development of the carposporic filaments, or *gonimoblasts*, into creeping threads which may be able to utilize food contained in neighbouring cells.

(2) *Cryptonemiales*.

Here there are definite pit connexions to the auxiliary cells, which serve not only for nutrition but also as starting points for the gonimoblast filaments. The auxiliary cells develop on special branches *before* fertilization and are actively concerned in the post-fertilization processes.

(3) *Gigartinales*.

A normal intercalary cell of the mother plant is set aside as an auxiliary cell *before* fertilization.

(4) *Rhodymeniales*.

The auxiliary cells are small, and though cut off *before* fertilization they only develop *after* that process has taken place.

(5) *Ceramiales*.

The auxiliary cell is cut off from a support cell *after* fertilization and as a direct consequence of the process. Series (3)–(5) should probably be regarded as examples of progressive reduction.

Auxiliary cells absent	(Nemalionales)	
	(Gelidiales)	No
Auxiliary cells present *before*	(Cryptonemiales)	procarp
fertilization	(Gigartinales)	
	(Rhodymeniales)	
Auxiliary cells develop *after*	(Ceramiales)	Procarp
fertilization		present

In 1926 Sjöstedt created two new orders, the Sphaerococcales and the Nemastomales, but in this volume the genera composing these two new orders are retained in the orders to which they have belonged in the past.

The antheridial plants, which are often paler in colour and more gelatinous, were first mentioned in a letter to Linnaeus in 1767 when they were considered to be male by analogy. The antheridia are either borne over the whole surface (e.g. *Dumontia*), or else in localized sori. These sori are reticulate in *Rhodymenia*, band-like in *Griffithsia*, borne on special branches in *Polysiphonia*, sunk in conceptacles in *Laurencia* and occur on the tips of the thallus in *Chondrus*. Very little is known about the seasonal periodicity of the male plants, which are often less frequent than either the female or tetrasporic plants, but this may be due purely to lack of observation, although it is also possible that the male plants are gradually becoming functionless. The antheridia often appear in an orderly sequence, being cut off usually as subterminal or lateral outgrowths from the antheridial mother cell. If they have been borne on a special part of the thallus (e.g. *Delesseria*) this may fall off or die away after fruiting is completed, whilst in other cases the mother cells simply revert to a normal vegetative state. The different types of male plant have been classified by Grubb (1925) as follows:

(*a*) The antheridial mother cell does not differ from the vegetative cells either in form or content, nor are the antheridia covered by a continuous outer envelope, e.g. *Nemalion, Batrachospermum*.

(*b*) The antheridial mother cells are differentiated from the vegetative cells, and the antheridia are surrounded by a common outer sheath, which is later pierced by holes or else gelatinizes in order to allow the ripe spermatia to escape:

(1) The antheridia develop terminally, e.g. *Melobesia, Holmsella*.
(2) The antheridia develop subterminally:

    (*a*) Two primary antheridia, e.g. *Delesseria sanguinea, Chondrus crispus*.
    (*b*) Two or three primary antheridia, e.g. *Scinaia furcellata, Lomentaria clavellosa*.
    (*c*) Three primary antheridia, e.g. *Ceramium rubrum, Griffithsia corallina*.
    (*d*) Four primary antheridia, e.g. *Polysiphonia violacea, Callithamnion roseum*.

The primary antheridia are commonly succeeded by a second crop which arises within the sheaths of the first, but a third crop only occurs in a few genera. In *Nemalion*, after the spermatium has become attached to the trichogyne, the nucleus undergoes a division but only one of the daughter nuclei acts as the fertilizing agent: this feature has led to the suggestion that in the more advanced red algae the contents of the antheridium are equivalent to a body which formerly did divide.

The tetraspores are either formed in superficial tetrasporangia or else they are sunk into the thallus, in which case the fertile branch often becomes swollen and irregular in outline, whilst in the genus *Plocamium* there are special lateral fertile branches or *stichidia*. Meiosis normally occurs at the formation of the tetra-spores, but when the spores develop on sexual haploid plants, as sometimes happens, there is no meiotic division and the products function as *monospores*. In *Agardhiella tenera* apospory is some-times found and again there is no meiosis so that a succession of asexual plants can occur. In the Nemalionales reproduction by means of monospores is quite common though the homologies of these bodies are somewhat uncertain. In some of the Eu-florideae (*Plumaria, Spermothamnion*) polyspores or paraspores develop on the diploid plants, but it has recently been shown that these are in some cases morphologically equivalent to tetraspores, whilst in others, e.g. *Plumaria*, they form the reproductive organs of a tri-ploid generation (cf. p. 238). Experimental cultures made on oyster shells have demonstrated that there are good grounds for believing that of the four spores in a tetrad two will give rise to female plants and two to male plants. Observations have been published showing that monospores, carpospores and tetraspores of some Rhodo-phyceae appear capable of a small degree of motion, the spores of the Bangiaceae being the most active among those investigated. The mechanism of this movement is not understood, and it is doubtful whether it is sufficient to give it any significance in the reproduc-tive processes of the plants.

Whilst there are apparently very few truly parasitic species among the Chlorophyceae and Phaeophyceae, nevertheless in the present group there are some very definite partial or total parasites. *Ceramium codicola* occurs on a Californian species of *Codium* and is said to be a partial parasite; *Ricardia Montagnei* is probably a total parasite at some stage of its existence, and the members of the two genera, *Janczewskia* and *Peysonelliopsis*, are probably entirely parasitic. In European waters *Choreonema, Schmitziella, Choreo-colax, Harveyella* and *Holmsella* are all to be regarded as partial or total parasites, and to this list *Polysiphonia fastigiata* should perhaps be added, since it is always found on one particular host, *Ascophyllum*. The order is principally marine, but there are a few fresh-water representatives, e.g. *Batrachospermum*, *Lemanea* and *Hildenbrandtia*, which are usually confined to fast-flowing streams where there is an abundance of aeration.

Proto-florideae

BANGIACEAE: *Porphyridium cruentum* (*porphyridium*, diminutive of purple dye; *cruentum*, blood red). Fig. 142.

This alga has had an extremely varied history, having been placed at various times in both the Palmellaceae and Schizogonia-ceae of the Chlorophyceae, near to *Aphanocapsa* in the Cyano-phyceae, and among the Bangiaceae in the Rhodophyceae where

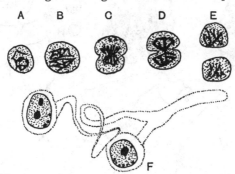

Fig. 142. *Porphyridium cruentum*. A–E, stages in nuclear and cell division (× 1280). F, cells connected by stalks after division (× 1280). (After Zirkle and Lewis.)

it finds a home at present. The single cells are united into a one-layered, gelatinous colony of a blood red colour which is found on the soil. Cell divisions take place in all directions, and when a cell divides the sheath elongates to form a kind of stalk which eventually ruptures. So far no form of sexual reproduction has been observed. In each cell there is one large chromatophore with cyanophycin granules around the periphery and also a central nuclear-like body, composed largely of anabaenin, which undergoes a primitive form of mitosis at cell division. Whether this alga represents a primitive form or else is a much-reduced type cannot at present be determined.

*BANGIACEAE: *Porphyra* (purple dye). Fig. 143.

This is a genus which has a very wide range as it extends in the northern hemisphere from 40° to 71° N. and in the southern from the Cape of Good Hope to 60° S. It has a variable seasonal periodicity in English waters where its presence is determined by the amount of water available, e.g. whether the site is subject to spray, together with the intensity of light and shade. The plant is

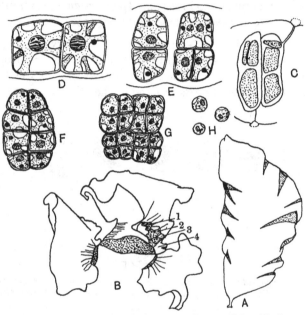

Fig. 143. *Porphyra.* A, thallus (× ⅓). B, attachment disk with three primary blades and four (1–4) secondary. C, formation of carpospores. D–H, formation of antherozoids in *P. tenera*. (A, D–H, after Ishikawa; B, C, after Grubb.)

flat and membranous, whilst in the common species, *P. umbilicalis*, there are a number of growth forms, the shape, width and length of the various forms being determined by the age of the plant, the height above mean sea-level and the type of locality. The plants are attached by means of a minute adhesive disk which is capable of producing lateral extensions from which new fronds may be proliferated. The disk is composed of long slender filaments together with some short stout ones, those near to or in actual contact with the substrate swelling up, branching and producing suckers or haptera which are apparently capable of penetrating dead wood or the tissue of brown fucoids. In the latter case there is evidently a capacity for epiphytism once contact is secured, and there is even some evidence of partial parasitism. In California, *P. naiadum* is an obligate epiphyte on *Phyllospadix* and *Zostera*, two marine phanerogams.

The gelatinous fronds of *Porphyra*, which are normally monostromatic although they become distromatic during reproduction, are composed of cells that possess stellate chromatophores with a pyrenoid, the process of nuclear division being intermediate between mitosis and amitosis. Reproduction is by means of monospores, carpogonia, which have rudimentary trichogynes, and antheridia, the carpogonial areas occupying a marginal position on the thallus. All the frond, except the basal region, can produce antheridia, but fertilization has never actually been observed although there is strong evidence which suggests that it does take place. The male thalli are paler in colour than the female, and each antheridial mother cell gives rise to sixty-four or 128 antheridial cells, each of which produces one spermatium. The fertilized carpogonium divides into four or eight cells that represent primitive carpospores; these are typically diploid whereas here they are haploid because a form of meiosis occurs when the fertilized carpogonium begins to divide. The carpospores eventually germinate to form a creeping filament, and it has recently been shown that spores from these threads are liberated and when germination has commenced it represents the commencement of a new *Porphyra* plant. It is suggested that the protonemal stage is equivalent to an adelophycean or dwarf generation in the life cycle, and further work on this part of the life history might produce interesting results.

The plant, which is called "lava" in England, "sloke" in Ireland and "slack" in Scotland, was formerly used as a food when it had been boiled and seasoned with spices and butter. It is still used as a food and medicine in Hawaii under the name of *Limu Luau*. In Japan, where there are over 2000 acres in cultivation, it is grown on bamboo bushes planted out between the tide marks where there is a depth of 10–15 ft. at high water. After collection, the plants are stirred in fresh water in order to cleanse them, chopped up into small bits, dried in the air and then pressed into sheets which, after crisping over a fire, can be dropped into culinary dishes in order to add a savour.

Eu-florideae

NEMALIONALES, GELIDIALES

*Batrachospermaceae: *Batrachospermum* (*batracho*, frog; *spermum*, seed). Fig. 144.

Two genera commonly found in fresh waters, *Batrachospermum* and *Lemanea*, belong to the Nemalionales. *Batrachospermum moniliforme*, which is a very variable species, is found attached to

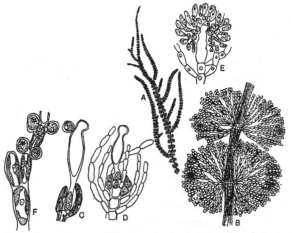

Fig. 144. *Batrachospermum moniliforme.* A, plant. B, portion of plant. C, carpogonial branch (× 480). D, fertilized carpogonium (× 360). E, mature cystocarp (× 240). F, antheridia (× 640). (A, B, after Oltmanns; C–F, after Kylin.)

stones in swift-flowing waters of the tropics and temperate regions. The thallus is soft, thick and gelatinous, the primary axis, which grows from an apical cell, being formed of a row of large cells. Numerous branches arise in whorls from the nodes, the basal regions of these branches producing corticating cells that grow downward and invest the main axis. The cells of the thallus are uninucleate and contain only one pyrenoid. Reproduction takes place by means of monospores, carpogonia and antheridia, the latter organs arising as small, round, colourless cells at the apices of short, clustered, lateral branches. The carpogonia are also terminal and possess a trichogyne which shrivels away after fertilization. The nucleus of the fertilized carpogonium divides twice, thus giving rise to four cells, and from these the short gonimoblast filaments grow out and finally terminate in a sporangium that produces a single naked carpospore which soon secretes a cell. A character of many of the Nemalionales is the occurrence of meiosis immediately after fertilization so that the carposporophyte is haploid as in the Proto-florideae.

The life history of the related genus *Nemalion* is similar to that of *Batrachospermum*, except that when the spermatia are liberated the nuclei are often in prophase, the division being completed when they have become attached to a trichogyne. This division has suggested to some workers that the spermatium is really homologous to an antheridium, but it might also be argued that it is a relic of a time when an antheridium produced more than one spermatium.

*CHAETANGIACEAE: *Scinaia* (after D. Scina). Fig. 145.

This is a widespread genus with its home primarily in the northern hemisphere, the commonest species, *S. furcellata*, being monoecious, although one may find monospores and spermatia on the same plant. The fronds, which arise from a discoid holdfast, are subgelatinous, cylindrical or compressed and dichotomously branched. The centre of the thallus is composed of both coarse and fine colourless filaments, the former arising from the apical cell and the latter from the corticating threads. There is also a peripheral zone of horizontal filaments that terminate in short corymbs of assimilatory hairs with a large colourless cell in the centre. These two types of epidermal cell are apparently differentiated near the apex of the thallus, the small ones giving rise to hairs, monosporangia or antheridia. The large colourless cell is said to form a protection against intense light, but it may also be a relic of a tissue which formerly had a function that has since been lost. One or two spores are formed in each monosporangium, whilst the spermatia arise in sori, forming bunches of cells at the ends of the small-celled assimilatory branches. The carpogonial branch is three-celled, the reproductive cell containing two nuclei, one in the carpogonium proper and one in the trichogyne. The second cell of the carpogonial branch gives rise to a group of four auxiliary cells which are rich in protoplasm, whilst the sterile envelope of the cystocarp arises from the third cell.

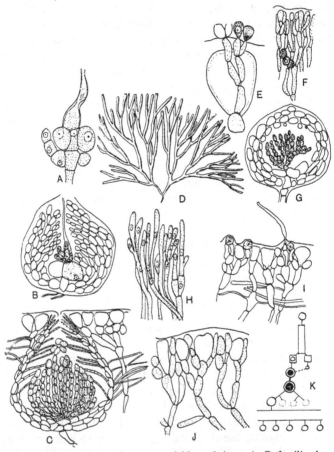

Fig. 145. *Scinaia furcellata*. A, carpogonial branch (× 700). B, fertilized carpogonium. C, cystocarp (× 195). D, plant (× ½). E, antheridia (× 700). F, young carpogonial branch (× 425). G, young cystocarp (× 232). H, undifferentiated threads at apex of thallus (× 425). I, monospores and a hair (× 340). J, differentiated cortex (× 429). K, life-cycle diagram. (C, after Setchell; D, original; rest after Svedelius.)

It is now certain that in the related genus *Chaetangium*, and probably also in *Galaxaura*, the wall of the cystocarp arises from the cell containing the fertilized nucleus, so that it is composed of fertile gonimoblasts and not sterile tissue. The fertilized nucleus in

Scinaia travels to the four auxiliary cells which have fused together and there meiosis ($n = 10$) occurs, after which one daughter nucleus passes back into the carpogonium and is concerned with the development of the gonimoblasts. There are, of course, no diploid plants because meiosis occurs immediately after fertilization.

CHAETANGIACEAE: *Liagora* (after one of the nereids). Fig. 146.

The principal interest of this genus, which is very similar morphologically to *Scinaia*, is provided by the species, *Liagora*

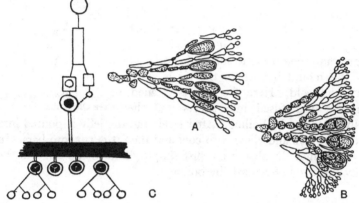

Fig. 146. *Liagora*. A, carpospores of *L. viscida* ($\times$ 320). B, carpospores in fours in *L. tetrasporifera* ($\times$ 320). C, life cycle of *L. tetrasporifera*. (A, B, after Kylin; C, after Svedelius.)

tetrasporifera, an inhabitant of the Canary Islands. The carpospores of this plant divide to give four spores which must probably be regarded as tetraspores, and although no cytological evidence is available, nevertheless it is presumed that meiosis is delayed to the time when the carpospores germinate. In this species, therefore, the carposporophyte is diploid, but at the same time no independent tetrasporic diploid generation develops. The remaining species of the genus behave like the other members of the Nemalionales, although in *L. viscida* the carpogonial branch is five-celled instead of the usual three cells.

GELIDIACEAE: *Gelidium* (congealed). Fig. 147.

In this genus there is no auxiliary cell, but the presence of the nutrient cells results in the production of a complex structure composed of several carpogonia together with nutrient cells, and more than one of these carpogonia may be fertilized. The genus is the principal source of agar-agar, a gelatinous medium much used in mycology and bacteriology, in the manufacture of size and in culinary operations. Agar-agar is manufactured primarily in Japan where it possesses various names, Kanten, Japanese, Bengal or Oriental isinglass, and Ceylon or Chinese moss. The plants contain about 76 % of the primary gelatinous material, gelose, and are dived for between May and October, after which they are allowed to dry and bleach in the open, and then they are sold to factories up in the mountains where the air is pure, dry and cold. Here the alga is cleaned, drained and fused into sheets and the jelly extracted by boiling. After straining, the jelly is poured into wooden trays and allowed to cool and then it is cut into bars. In former times the algae were just simply dried in the sun and the jelly extracted afterwards by boiling.

Fig. 147. *Gelidium corneum.* (After Oltmanns.)

CRYPTONEMIALES

Dumontiaceae: Dudresnaya (after Dudresnay de St-Pol-de-Léon). Fig. 148.

The cylindrical, much-branched thallus is soft and gelatinous, consisting when young of a simple articulated filamentous axis with whorls of dichotomously branched ramuli, although in older plants the central axis becomes polysiphonous and clothed with densely set whorls of branches. The plants are dioecious, the males being somewhat smaller, paler and fewer in number than the females. The carpogonial branches of *D. coccinea* arise from the lower cells of short side branches and when fully developed are composed of seven to nine cells: they are branched once or twice and may have short sterile side branches arising from the lowest cell. In the middle of the mature carpogonial branch there are two to three larger cells which function in a purely nutritive capacity, whilst the auxiliary cells develop in similar positions on neighbouring branches that are homologous with the carpogonial branches. After fertilization the carpogonium sends down a protuberance containing the diploid nucleus and this cuts off two cells when it is near to the nutrient cells of the carpogonial branch. These all fuse together and sporogenous threads, each carrying a diploid nucleus, then grow out towards the auxiliary cells on the other branches. When

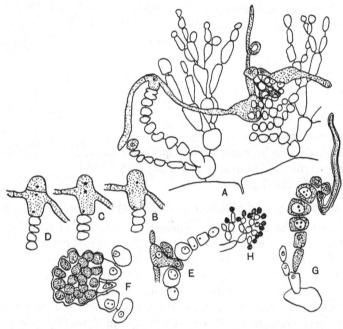

Fig. 148. *Dudresnaya.* A–D, stages in development of cystocarp, *D. purpurifera.* E, F, stages in development of cystocarp in *D. coccinea* after fertilization (× 486). G, *D. coccinea*, carpogonial branch (× 486). H, *D. coccinea*, antheridia (× 510). (A–D, after Oltmanns; E–G, H, after Kylin.)

these filaments fuse with an auxiliary cell the latter forms a protuberance into which the diploid nucleus passes, and after this has divided once the protuberance containing one of the daughter nuclei is cut off by a wall. The gonimoblast filaments then grow out as a branched mass from this protuberance of the auxiliary cell. Each sporogenous thread sent out from the original fusion cell may unite with more than one auxiliary cell in the course of its wanderings through the thallus, so that one fertilization may result in the production of a number of carposporophyte generations.

SQUAMARIACEAE: *Hildenbrandtia* (after F. E. Hildenbrandt). Fig. 149.

This genus is characteristic of a small group of red algae all of which form thin crusts on stones or other algae, and it is frequently difficult to distinguish in the field from similar encrusting brown types such as *Ralfsia*. The frond is horizontally expanded into a thin encrusting layer composed of several layers of cells arranged in vertical rows, the plants forming indefinite patches that are attached by a strongly adhering lower surface. The genus is both marine and fresh water, *Hildenbrandtia rivularis* appearing frequently in rivers and streams. The principal mode of reproduction is by means of tetraspores which are produced in sporangia borne in rounded or oval conceptacles that are sunk in the thallus.

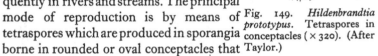

Fig. 149. *Hildenbrandtia prototypus*. Tetraspores in conceptacles (× 320). (After Taylor.)

*CORALLINACEAE: *Epilithon* (*epi*, above, *lithon*, stone). Fig. 150.

This and the succeeding type belong to the Corallinaceae, a family of calcareous red algae which have played much part in the building up of rocks and coral reefs and which have been known as fossils from the earliest geological strata. The present type has been selected because the common species, *E. membranaceum*, is less calcified than other members of the Corallinaceae and thus forms very convenient material for sectioning and demonstration purposes without the trouble of decalcification. The thallus, which forms a crust on other algae or phanerogams, consists of a single cell layer composed of large cells, from each of which is cut off a small upper cell that goes to form the outer lime-encrusted layer.

Further divisions take place internally from the large basal cells so that one finally obtains rows of erect filaments growing side by side. The various reproductive organs are borne in conceptacles on separate plants; in the male plants, for example, there are a number of two-celled filaments in the centre of every conceptacle. The basal cells of these threads cut off two antheridial mother cells which in their turn produce two antheridia, whilst the upper cells grow out to form the walls of the conceptacle. In the female plant the central threads form three-celled carpogonial branches, whilst the outer threads develop into two-celled filaments that are modified auxiliary cells. After fertilization the carpogonium and the cell below it fuse together and send out a filament to the lower cell of the auxiliary branch. Finally, all the auxiliary and nutritive cells fuse to give one long fusion cell from which very short gonimoblast filaments grow out. In the tetrasporic plant there are

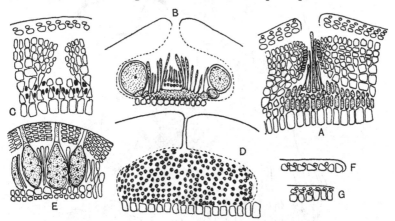

Fig. 150. *Epilithon membranaceum.* A, carpogonia (× 360). B, conceptacle with ripe carpospores (× 240). C, young antheridial conceptacle (× 510). D, mature antheridial conceptacle (× 426). E, tetraspores (× 228). F, G, thallus construction (× 360). (After Kylin.)

simple filaments which give rise to the tetrasporangia and branched sterile filaments that form the roof to the conceptacle by the process of division and elongation, the original roof being cast off: finally, a pore develops above each group of tetraspores.

*CORALLINACEAE: *Corallina* (coral). Fig. 151.

Both this and the preceding genus are examples of the "fountain" type of construction (cf. p. 212). In *Epilithon* the original construction has been much modified because of its habit, but it can be observed extremely well in *Corallina*. The erect plants, which are jointed, cylindrical or compressed, arise from calcified encrusting basal disks or prostrate interlaced filaments. Branching, which is frequent, is either pinnate or dichotomous. There is a central core of dichotomously branched filaments with oblique filaments growing out at the swollen internodes to form a cortical layer, the whole being covered by a dense coating of lime, whilst in *C. rubens* there may also be epidermal hyaline hairs. The plants are monoecious or dioecious, the reproductive organs being borne in terminal or lateral conceptacles. The carpogonia, which are not

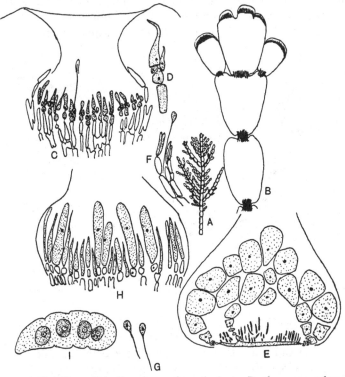

Fig. 151. *Corallina officinalis*. A, portion of plant. B, the same enlarged. C, carpogonial conceptacle (× 210). D, single carpogonial branch (× 342). E, fusion cell, gonimoblasts and carpospores (× 120). F, development of antheridia (× 420). G, mature spermatia (× 648). H, young tetrasporic conceptacle (× 240). I, mature tetraspore (× 270). (A, B, after Oltmanns; rest after Suneson.)

calcified, arise from a kind of prismatic disk formed from the terminal cells, these cells also functioning later as the auxiliary cells. As a result of oblique divisions, one to three embryo carpogonial branches are formed on each mother cell, but only one of these finally develops into the mature two-celled carpogonial branch with its long trichogyne. After fertilization a long or rounded fusion cell is formed by the auxiliary cells, and this contains both fertilized and unfertilized carpogonial nuclei. The antheridia are much elongated, and after liberation the spermatia round off and remain attached to the antheridial wall by means of a long thin pedicel in *C. officinalis* and by a short stalk in *C. rubens*.

CERAMIALES

DELESSERIACEAE: *Delesseria* (after Baron Delessert). Fig. 152.

The large, thin, leafy fronds, which are bright red in colour,

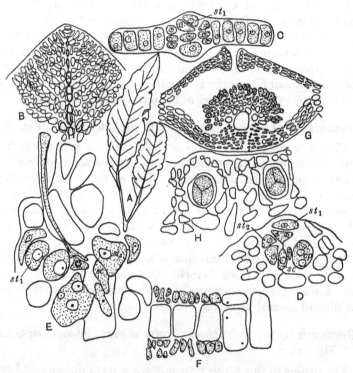

Fig. 152. *Delesseria sanguinea*. A, plant. B, apex of thallus to show cell arrangement (×258). C, first stage in formation of carpogonial branch. st_1 = first group of sterile cells (×408). D, later stage of same. cp = carpogonial branch, sc = support cell, st_1 = first, and st_2 = second group of sterile cells (×408). E, mature carpogonial branch. sc = support cell. st_1 = first sterile branch, st_2 = second sterile branch (×720). F, formation of antheridia in related genus, *Nitophyllum*. G, transverse section of mature cystocarp in the related genus *Nitophyllum*. H, tetraspores (×360). (A, F, G, after Tilden; B–D, after Kylin; E, H, after Svedelius.)

possess a very conspicuous mid-rib with both macro- and micro-scopic veins and they form magnificent plants for pressing as herbarium specimens. The complex nature of the laciniate or branched thallus can be seen from the figure. There are three orders of cells with considerable intercalary division, although the cortication of the primary cell filaments to form the veins does not involve intercalary division. The cells of the thallus also become united by means of secondary protoplasmic threads and they may also develop thin rhizoids. The cystocarps are small stalked bodies which are borne on the mid-rib, whilst the tetrasporangia are produced in special fertile leaflets that arise from the mid-rib, but as these do not possess the power of intercalary growth they differ slightly in structure from the vegetative thallus. In the related genus *Martensia* each tetraspore mother cell is multinucleate, containing about fifty nuclei all of which degenerate except for one, and from this the four nuclei of the tetraspores are produced.

RHODOMELACEAE: *Janczewskia* (after E. de Janczewski). Fig. 153.

This is a remarkable hemi- or holo-parasitic genus which is always to be found on other members (*Laurencia*, *Chondria* and *Cladhymenia*) of the same family. One of the most interesting features of this parasitism is that the genus is very closely related to *Laurencia* and yet is parasitic upon various species of that genus. All the species have organs of contact or penetration, the latter being fungal-like filaments which establish pit connexions with the cells of the host. Each individual plant is a coalescent tubercular mass composed of fused branches that grow from an apical cell buried in a pit as in *Laurencia*. The sexual plants are dioecious and the diploid asexual plant also occurs.

*RHODOMELACEAE: *Polysiphonia* (*poly*, many; *siphonia*, siphons). Fig. 154.

The thallus in this genus generally arises from decumbent basal filaments that are attached to the substrate by means of small flattened disks. Many species are epiphytic on other algae whilst *P. fastigiata*, which is always found on the fronds of the fucoid *Ascophyllum nodosum*, is probably a hemi-parasite. The thallus is laterally or dichotomously branched and bears numerous branches which are shed annually in the perennial forms before winter and are re-developed in the spring. The main axes and branches are corticate or ecorticate, and possess a polysiphonous appearance due to the single axial cell series being surrounded by four to twenty-four pericentral cells or siphons. The corticating cells, when present, are always shorter and smaller and are often only found in the basal portions of the stem. The ultimate branches are not

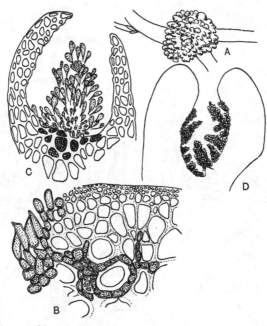

Fig. 153. *Janczewskia.* A, *J. moriformis* on *Chondria* sp. (×6). B, filaments of *J. lappacea* in host, *Chondria nidifica* (×180). C, longitudinal section of cystocarp of *J. moriformis* (×180). D, antheridial conceptacle of *J. lappacea* (×180). (After Setchell.)

polysiphonous and frequently terminate in delicate multicellular hairs.

The colourless antheridia, which are formed in clusters, are borne on a short stalk that morphologically is a rudimentary hair. In *Polysiphonia violacea*, where the haploid number of chromosomes is twenty, the two basal cells of the hair are sterile, the upper one giving rise to a fertile polysiphonous branch and a sterile hair. One or more mother cells are formed from all the pericentral cells on the fertile branch, and each mother cell produces four antheridia in two opposite and decussate pairs, the first and third appearing before the second and fourth. There is no secondary crop in this species. The carpogonial branches are also formed from hair

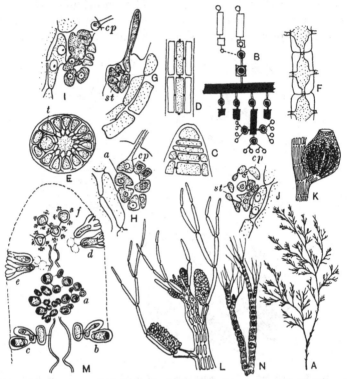

Fig. 154. *Polysiphonia violacea.* A, plant of *P. nigrescens* (× ⅓). B, life cycle. C, apex and cells cut off from central cells. D, thallus construction in longitudinal section. E, transverse section of thallus, *P. fastigiatum.* *t* = young tetraspore. F, protoplasmic connections of axial thread. G–J, stages in development of carpospores. *cp* = carpogonium, *a* = auxiliary cell, *g* = gonimoblast, *st* = sterile cells (× 400, J × 260). K, cystocarp of *P. nigrescens* with ripe carpospores (× 33). L, antheridial branch (× 35). M, *a–f*, stages in development of antheridia. N, *P. nigrescens*, tetraspores (× 33). (A, K, N, after Newton; B, after Svedelius; C, F, schematic; D, E, after Oltmanns; G–J, after Kylin; L, after Grubb; M, after Tilden.)

rudiments, the support cell cutting off a small section from which lateral sterile cells arise. Later on a fertile pericentral cell is cut off, and this gives rise to the four-celled carpogonial branch, the carpogonium being of interest because there is also a persistent nucleus in the trichogyne.

After fertilization has taken place the auxiliary cell is cut off from the apex of the fertile pericentral cell and in addition two branch systems composed of nutrient cells appear. When the zygote nucleus has divided the two daughter nuclei (only one of the two in *P. nigrescens*) pass into the auxiliary cell which has become fused to the carpogonium in the meantime, and there the two nuclei are isolated from the carpogonium by a new wall. By this time the carpogonium and its three lower cells have broken down. The auxiliary cell then fuses with the pericentral cell and after the two diploid nuclei have passed into it, it unites with the other support and axial cells to give a large fusion cell. The diploid nuclei undergo a number of divisions and the products pass into lobes that are budded off from the fusion cell. Each lobe then gives rise to a two-celled gonimoblast filament, the first cell acting as a stalk cell whilst the end cell produces a carpospore. The wall of the cystocarp is two-layered, the outer wall being formed from the lateral sterile cells that are cut off from the support cell, whilst the inner lining is formed from the axial cell of the fertile segment. The tetrasporangia, which develop from pericentral cells, are protected by being embedded in the thallus, a feature which results in the fertile branch usually being much swollen and distorted.

CERAMIACEAE: *Griffithsia* (after Mrs Griffiths). Fig. 155.

The monosiphonous ecorticate fronds are composed of large multinucleate cells connected to each other by a pore, although this is often closed by a plug. In *G. globulifera* the larger cells may each have as many as 3000–4000 nuclei. Vegetative division is brought about either by the cutting off of terminal segments from the end cells or else by the delimitation of a small cell from the upper edge, but as this grows very rapidly by mere swelling the appearance of a false dichotomy is produced. In *G. corallina* miniature shoots and also delicate colourless branched hairs develop from the large cells of the main thallus. Regeneration can occur in order to replace an old cell or one that has been wounded, the process involving the two neighbouring cells which send out tubes that meet and fuse. The sessile antheridia are borne on the distal ends of much-branched dwarf shoots which surround the nodes of the main thallus in tufts or dense whorls, each branch arising as a protuberance that is cut off from one of the large axial

cells. The primary cell of a carpogonial branch, which is cut off
from the apex of a growing cell, becomes pushed down to the side

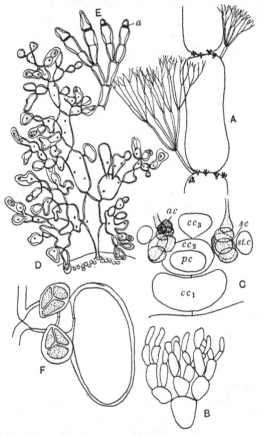

Fig. 155. *Griffithsia corallina*. A, portion of plant with short shoots and
branched hairs (× 18). B, short shoot magnified (× 312). C, carpogonial branch
(× 370). *ac* = auxiliary cell, *cc₁*, *cc₂*, *cc₃* = central cells, *pc* = pericentral cell,
sc = support cell, *stc* = sterile cell. D, antheridial branch (× 720). E, plant with
antheridia (*a*) (× 3·2). F, tetrasporic branch (× 222). (A–C, F, after Kylin;
D, E, after Grubb.)

where it divides into three cells. The second cell forms the fertile
central cell and gives rise to three pericentral cells, one of which
produces a one-celled branch whilst the others produce two-celled
branches. The basal cell of each of these two-celled branches gives

rise to a four-celled carpogonial branch. In the original branch of three cells the first cell gives rise to a protective branch after fruiting has occurred, whilst the third cell remains sterile throughout. The tetraspores, which are borne in whorls, are partly covered by involucral cells. At tetraspore formation, after a small support cell has been cut off from an ordinary vegetative cell it proceeds to cut off several side cells, each of which functions as a tetrasporangium. Finally the support cell cuts off two sterile cells at its apex, the distal one enlarging to become a protective cell for the whorl of tetraspores.

*CERAMIACEAE: *Callithamnion* (*calli*, beauty; *thamnion*, small bush). Fig. 156.

This is a genus of very beautiful and delicate plants that possess filamentous branched fronds which are either monosiphonous or else corticated at the base, the cortication being formed by rhizoidal filaments. The cells of the vegetative thallus are multinucleate, and in *C. byssoides* there are protoplasmic pseudopodia projecting internally from the ends of the cells, and although these strands are apparently capable of some movement their function is obscure.

The antheridia, which form hemispherical or ellipsoidal tufts on the branches, arise as lateral appendages, the first cell to be cut off being the stalk cell. This stalk cell gives rise to a group of secondary cells which later on divide to form branches composed of two to three cells, each terminating in an antheridial mother cell. In this genus there may be two or even three crops of antheridia arising successively in the same place, each mother cell producing about three antheridia in every crop. The cystocarps, which are usually present in pairs and enclosed in a gelatinous envelope, arise as follows. Two cells are cut off from a cell in the middle of a branch and these function as the auxiliary *mother* cells. From one of them the four-celled carpogonial branch is produced, whilst after fertilization both auxiliary mother cells divide and cut off a small basal cell. The fertilized carpogonium also divides into two large cells, each of which cuts off a small sporogenous cell that fuses with the adjacent auxiliary cell. As a result of this fusion each auxiliary cell can receive a diploid nucleus which soon after its entry divides into two; one daughter nucleus passes to the apex of the auxiliary cell, whilst the other, together with the nucleus of the auxiliary cell, is cut off by a wall. It is from the large upper cell that the gonimo-blast filaments arise and so the mature cystocarp is produced.

The sessile tetrasporangia arise in acropetal succession as lateral outgrowths of the vegetative cells of young branches. In *C. bra-chiatum* mature tetrasporangia and antheridia have been found on the same plant, whilst other plants have been reported that bear

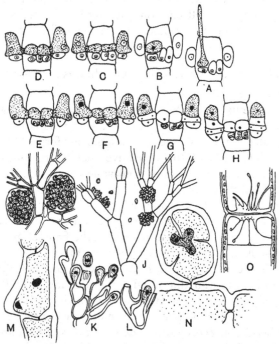

Fig. 156. *Callithamnion*. A–I, stages in development of carpospores after fertilization. J, antheridia. K, the same enlarged. L, secondary spermatium. M, young tetraspore. N, mature tetraspore. O, amoeboid processes. (A–I, after Oltmanns; J–L, after Grubb; M, N, schematic after Mathias; O, after Phillips.)

both tetrasporangia and cystocarps. In these cases the nuclei of the carpospores were found to be haploid whilst those of the vegetative cells were diploid, so that if fertilization occurred there must have been two meiotic divisions, one before and one after fertilization. If only one meiotic division occurs then it must be supposed that the carpospores arose apogamously. *Spermothamnion Turneri* is another plant in which sex organs have also been reported on normal tetrasporic plants, but as the procarp branch in this case develops normally without meiosis the carpogonium is diploid. Fusion of the nuclei in the carpogonium has been observed so that the gonimoblast filaments must be tetraploid, but unfortunately the fate of the carpospores is not known. In *S. Snyderae* the tetra-sporangia are replaced by *polysporangia* which must be regarded as homologous structures. The mother cells of each polysporangium contain two to nine nuclei and they give rise to twelve, sixteen, twenty, twenty-four or twenty-eight spores.

CERAMIACEAE: *Plumaria* (*pluma*, soft feather). Fig. 157.

The filamentous thallus is much branched, the main axis, which is monosiphonous throughout, being ecorticate near the apex but

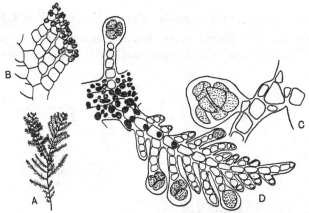

Fig. 157. *Plumaria elegans*. A, plant (× ⅔). B, antheridial ramuli (× 180). C, paraspores (× 213). D, tetrasporic ramuli (× 126). (A, original; B, after Drew; C, D, after Suneson.)

corticate below. The antheridia are borne on special branches, whilst the four-celled carpogonial branch develops from the sub-terminal cell of an ordinary branch. In northern waters *P. elegans* never bears sex organs and only plants with paraspores are to be found, whilst in southern waters the sexual ($n = 31$) and tetrasporic plants ($n = 62$) are predominant. Recent investigation has shown that in this species we are concerned with a triploid race ($n = 93$) in the northern waters which reproduces by means of paraspores. There is apparently no relation between the triploid plants and the

238 RHODOPHYCEAE

other two races, and, furthermore, the triploid has the wider distribution because it is able to penetrate into the colder waters of the
north. Tetraspores are to be found on the triploid plants but their
chromosomal complement and fate are not known. Although both
tetra- and parasporangia arise from a single cell it is doubtful if the
two structures are homologous. The reasons for this are first, the
difference in chromosomal complement, secondly, the absence of
any apparent relationship with the haploid and diploid plants, and
thirdly, differences in the mode of development of the para- and
tetrasporangia. This is the first cytological record of triploid plants
in the algae. Paraspores are also known in the related genus
Ceramium but their cytology, and hence homologies, are not known.

GIGARTINALES

CHOREOCOLACACEAE: *Harveyella* (after G. Harvey). Fig. 158.

This and the closely allied genus *Holmsella* are monotypic
genera each containing a holo-parasitic species, whilst *Choreocolax*

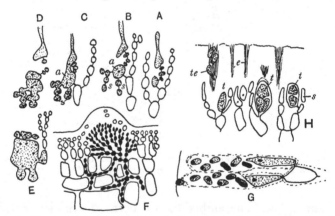

Fig. 158. *Harveyella* and *Holmsella*. A–E, stages in development of gonimoblasts after fertilization in *Harveyella mirabilis*. *a*=auxiliary cell, *s*=sterile
filaments. F, filaments of parasite, *Holmsella pachyderma*, in host. G, antheridia
of *Harveyella mirabilis*. H, tetraspores in *Holmsella pachyderma*. *e*=tracks left
after tetraspores have escaped. *s*=sterile cells, *t*=tetraspores in various stages,
te=escaping tetraspores. (After Sturch.)

is another parasitic genus very nearly related to them. *Harveyella
mirabilis* is parasitic on species of *Rhodomela* whilst *Holmsella
pachyderma* parasitises *Gracilaria confervoides*. They have little or no

colour of their own as might be suspected from their parasitic nature, and they send out branched filaments or haustoria into the tissues of the host. The parasites appear as external cushions lying on the branches of the host, each cushion, which is surrounded by an outer gelatinous coat, consisting of a central area that is four to five cells thick. In *Holmsella* the carpogonial branch is two-celled whilst in *Harveyella* it is four-celled, this feature forming one of the principal differences between them. The antheridial, carpogonial and tetrasporic plants are all separate, and the species are said to pass through the full floridean life cycle twice every year. It is clear that their much-reduced morphological features are to be associated with the parasitic habit, and have probably arisen as a result of the adoption of parasitism.

GIGARTINACEAE: *Chondrus* (cartilage). Fig. 159.

This is a widespread genus, many of the species appearing as a number of varieties, some of which are probably only ecological

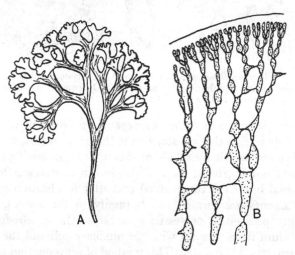

Fig. 159. *Chondrus crispus*. A, plant (× ⅔). B, transverse section of thallus (× 344). (A, after Newton; B, after Kylin.)

forms. *Chondrus crispus*, which is known as "Irish moss", contains 80% of water together with salts that control gelatinization. The plants are often collected and bleached, and then an extract is obtained which can be used in the curing of leather and the manufacture of size, and also for puddings and medicinal purposes.

*GIGARTINACEAE: *Phyllophora* (*phyllo*, leaf; *phora*, bear). Figs. 160, 161.

The stipitate fronds expand upwards into a rigid or membranous flat lamina which is either simple or divided, whilst proliferations may also arise from the margin or basal disks. Morphologically the thallus is composed of oblong polygonal cells in the centre bounded on the outside by cortical layers of minute, vertically seriate assimilatory cells. In some species secondary tissue has been observed near the axils of branches or at the base of the frond. The

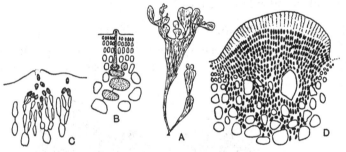

Fig. 160. *Phyllophora Brodiaei.* A, plant (× ⅓). B, carpogonial branch (× 250). C, transverse section of antheridial thallus (× 450). D, nemathecia with tetraspores (× 125). (A, original; B–D, after Kylin.)

plants are dioecious and the sex organs are borne in cavities in small fertile leaflets that are attached to the main thallus, the carpogonial leaflets, which are sessile or shortly stalked, arising laterally from the stipitate part of the main blade. In *P. membranifolia* the carpogonial branch is three-celled and after fertilization gonimoblast filaments are formed which ramify in the tissues, finally producing pedicellate or sessile cystocarps. In *P. Brodiaei* the carpogonium fuses directly with the auxiliary cell and the carposporic generation is omitted. This method of reproduction must be regarded as reduction from the ordinary process in so far as the usual rhodophycean life cycle is concerned. The tetraspores are borne in moniliform chains packed into wart-like excrescences or *nemathecia* which are borne on the female sexual plant. In *P. Brodiaei* the absence of carpospores led earlier investigators to regard the nemathecia as belonging to a parasitic plant, which in this case was given the name of *Actinococcus subcutaneus*, but it has since been shown that we are really dealing with a parasitic diploid generation.

In the related genus *Ahnfeldtia*, although reduction of the life cycle has gone still further, nevertheless nemathecia still appear and these also were formerly regarded as a parasite to which the name *Sterrocolax decipiens* was given. In this genus, however, there is

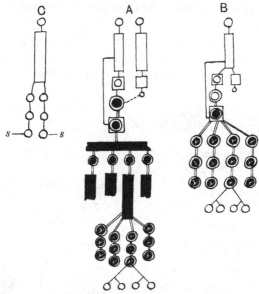

Fig. 161. Life cycles. A, *Phyllophora membranifolia*. B, *P. Brodiaei*. C, *Ahnfeldtia plicata*. s=monospores. (After Svedelius.)

neither fertilization nor meiosis and only degenerate procarps are formed; instead the nemathecia contain monospores that develop as follows. The warts, which arise as small cushions from superficial cells of the thallus, contain some cells that become flask-shaped together with other cells possessing denser contents that arise in groups at the upper ends of the filaments. These latter, which probably represent degenerate carpogonia, form the generative cells and they give rise to secondary nemathecial filaments, the apical cells of which function as the monosporangia. In *Ahnfeldtia*, therefore, the sporophytic generation has been completely suppressed, and this modified life cycle should be compared with that of *Lomentaria rosea* (cf. below) in European waters where the gametophytic generation has been secondarily suppressed. The monospores have been interpreted as morphologically equivalent to either the carpospores or the tetraspores, the latter interpretation being the one adopted in this volume.

RHODYMENIALES

*RHODYMENIACEAE: *Lomentaria* (pod with constricted joints). Fig. 162.

The filamentous fronds are hollow with constrictions at the nodes, whilst branching is irregular or unilateral. The hollow central region originates from a branching structure which later on

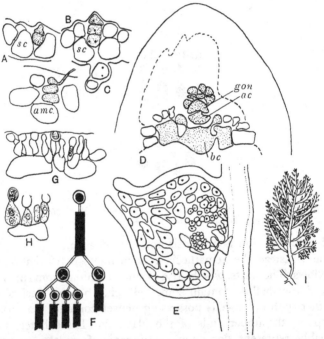

Fig. 162. *Lomentaria clavellosa.* A–C, development of carpogonial branch (×660). *amc*=accessory mother cell, *sc*=support cell. D, young cystocarp (×312). *ac*=accessory cell, *bc*=support cell, *gon*=gonimoblast. E, mature cystocarp (×90). F, *L. rosea*, life cycle. G, H, *L. clavellosa*, antheridia (×660). I, *L. clavellosa*, plant (×⅔). (A–C, F–H, after Svedelius; D, E, after Kylin; I, original.)

separates in order to form the outer cell layers, although a few longitudinal filaments are left in the centre. The plant, which is enclosed in a thick gelatinous cuticle, may bear unicellular hairs that have arisen from the epidermal layer. The adult thallus has developed from a group of eight to twelve apical cells, each of which produces a longitudinal filament, whilst the corticating threads develop from lateral cells which are cut off from each segment just behind the apex. The male plants, which are rare in nature, bear the antheridial sori on the upper regions where they form whitish patches. A system of branching threads, which appears as a preliminary to sorus formation, arises from a single central cell, and from each of these branching threads two to three antheridial mother cells grow out and increase in length. Depending on the species one, two or three primary antheridia arise from each mother cell and they may be followed by a crop of secondary antheridia. The procarp consists of a support cell with a three-celled carpogonial branch, both these and the antheridial mother cells being uninucleate, although the mature vegetative cells are multinucleate. There are one or two auxiliary cells, and after fertilization one of these receives a process from the carpogonium which carries with it the diploid nucleus. This auxiliary cell then proceeds to cut off a segment on the outer side, and from this a group of cells develops that ultimately gives rise to the gonimoblasts. The ripe cystocarps are sessile on the thallus and possess a basal placenta. The tetrasporangia are borne on the diploid plants in small cavities produced by the infolding of the cortex. In European waters *L. rosea*, which has a diploid chromosome number of twenty, is only known to produce tetraspores which apparently arise without undergoing meiosis. Individual spores germinate to give a new plant or else a whole tetrad may germinate to give a new plant. In *L. rosea*, therefore, the gametophytic generation is wholly suppressed and we have a diplont which behaves as a haplobiont in respect of its life cycle. In Pacific waters, on the other hand, the records suggest that the species behaves normally, whilst the other common species, *L. clavellosa*, also behaves in the normal fashion.

REFERENCES

Plumaria. BAKER, K. M. (1939). *Ann. Bot., Lond.,* N.S. **3**, 347.

Nemalion. CLELAND, R. E. (1919). *Ann. Bot., Lond.,* **33**, 323.

Porphyra. GRUBB, V. M. (1924). *Rev. Alg.* **3**, 1.

General. GRUBB, V. M. (1925). *J. Linn. Soc. (Bot.)* **47**, 177.

Porphyra. ISHIKAWA, M. (1921). *Bot. Mag., Tokyo,* **35**, 206.

Griffithsia. KYLIN, H. (1916). *Z. Bot.* **8**, 97.

Batrachospermum. KYLIN, H. (1917). *Ber. dtsch. bot. Ges.* **35**, 155.

Griffithsia. LEWIS, I. F. (1909). *Ann. Bot., Lond.,* **23**, 639.

Porphyridium. LEWIS, I. F. and ZIRKLE, C. (1920). *Amer. J. Bot.* **7**, 333.

Porphyra. MANGEOT, G. (1924). *Rev. Alg.* **1**, 376.

Callithamnion. MATHIAS, W. T. (1928). *Publ. Hart. Bot. Lab.* no. 5, p. 1.

Harveyella. STURCH, H. H. (1924). *Ann. Bot., Lond.,* **38**, 27.

Corallina. SUNESON, S. (1937). *Lunds Univ. Årsskr.* **33**, 1.

Delesseria. SVEDELIUS, N. (1911, 1912, 1914). *Svensk bot. Tidskr.* **5**, 260; **6**, 239; **8**, 1.

Scinaia. SVEDELIUS, N. (1915). *Nova Acta Soc. Sci. Upsal.* **4**, 1.

General. SVEDELIUS, N. (1931). *Beih. bot. Zbl.* **48**, 38.

Lomentaria. SVEDELIUS, N. (1937). *Sym. Bot. Upsal.* **2**, 1.

Polysiphonia. YAMANOUCHI, S. (1906). *Bot. Gaz.* **41**, 425; **42**, 401.

Corallina. YAMANOUCHI, S. (1921). *Bot. Gaz.* **72**, 90.

Porphyra. REES, T. K. (1940). *J. Ecol.* **28**, 429.

CHAPTER IX

REPRODUCTION, EVOLUTION AND FOSSIL FORMS

*REPRODUCTION

In this chapter it is proposed to give a general review of the various reproduction cycles that are to be found in the three principal algal groups, Chlorophyceae, Phaeophyceae and Rhodophyceae. It will also be instructive to ascertain whether such a survey can lead one to any helpful conclusions in considering evolution among and in the different groups. A study of fossil forms is of fundamental importance in any evolutionary or phylogenetic survey, but it must be clearly understood, however, that as the fossil forms of algae are largely confined to certain calcareous genera it is very difficult to draw any decisive conclusions. As a result, hypotheses must be based almost wholly upon living forms and these may have advanced far from their primitive ancestors, and furthermore, evolution may have proceeded at varying rates along the different lines. For this reason the bulk of the material set out in this chapter *can only be speculative, and students would do well to bear this in mind.* The necessity of basing hypotheses upon living forms also leads to the further complication that different authors inevitably propound schemes, and these may differ widely in representing their views of the lines along which the present living species have evolved. Here again it cannot be too strongly impressed upon the student that much of what follows must be attributed to the author's personal opinions, and these are not necessarily shared by other workers. The student should read the additional literature critically and then attempt to work out his own conclusions, and in this connexion it will often be found very helpful to draw up some form of schematic diagram.

As an essential preliminary it is convenient to recapitulate the principal life cycles to be found in the three groups, pointing out at the same time any problems that may arise immediately from such a survey. The life cycles of representative genera in the Phaeophyceae are shown in fig. 163, and a study of these enables one to make the following generalizations:

(1) The life cycle is by no means simple in most of these types and it frequently has no fixed relation to the nuclear cycle or to the cycle of reproductive bodies, and so it has been suggested that the term "life cycle" should be abandoned and replaced by the term "race cycle" because that indicates more clearly the numerous possible variations in the life history of any one species. Lying

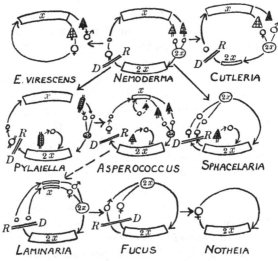

Fig. 163. Types of life cycle in the Phaeophyceae and their possible inter-relationships. RD = position of reduction division in the life cycle.

behind the race cycle is the fundamental nuclear cycle, but this is often obscured by the frequent repetition of any one generation. Whether these variations in the life history of any one species, e.g. *Ectocarpus siliculosus*, are to be related to differences in environment or whether they are due to genetical differences is a problem that still awaits solution.

(2a) Any thallus in the Ectocarpales, whether it be haploid or diploid, can produce an unlimited series of the same generation by means of zooids from plurilocular sporangia. In this connexion it is extremely instructive to compare and classify the Phaeophyceae in relation to the two types of sporangia. In Table II it will be seen that one can distinguish two primary divisions if one regards the antheridia and oogonia as modified plurilocular sporangia. This concept is inevitably bound up with the phylogeny of the Phaeophyceae because one can either read them as a series commencing with the undifferentiated plurilocular gametangia of the Ectocarpales, or else one can regard these structures as reduced antheridia and oogonia in which differentiation has been completely lost.

(2*b*) The presence of a unilocular sporangium always indicates the presence of a diploid thallus, and it invariably gives rise to haploid zooids.

TABLE II

I. One kind of plurilocular sporangium.

    (1) Uni- and plurilocular sporangia on the same individuals, e.g. *Ectocarpus* spp.

    (2) Uni- and plurilocular sporangia on different individuals, e.g. *Sphacelaria bipinnata*, *Cladostephus*.

II. Two kinds of plurilocular sporangia.

    (1) Meio- and megasporangia, e.g. *E. virescens*.

    (2) Antheridia and female gametangia (=plurilocular sporangia).

        (*a*) Unilocular sporangia on separate plants, e.g. *Sphacelaria hystrix*, *Halopteris filicina*.

        (*b*) Unilocular and both gametangia all on separate plants, e.g. *Sphacelaria Harveyana*.

    (3) Antheridia and oogonia (=plurilocular sporangia).
        Unilocular sporangia on separate plants, e.g. *Dictyota*, *Laminaria*.

    (4) Antheridia and oogonia representing modified micro- and megasporangia (=plurilocular sporangia).
        Unilocular sporangia on same plant, e.g. Fucales.

(2*c*) A haploid zooid, irrespective of the nature of the structure in which it was produced, can behave either as a gamete or as an asexual zooid.

(3) In many of the types it cannot be said that there is a regular alternation of cytological or morphological generations, even though it is potentially possible. Although by no means entirely satisfactory, in a good many cases the race cycle can perhaps be best described as possessing an irregular alternation of generations.

(4) Theoretically it is obvious that there are three possibilities which can be suggested in order to explain the origin and development of the Phaeophyceae:

A. Plants that are haploid throughout their life cycle, except for the zygote, represent the primitive condition, and the diploid stage became interpolated by a gradual delay in the occurrence of meiosis. Against this possibility it may be pointed out that

(i) There are very few Phaeophyceae in which the haploid generation is wholly dominant. It is possible, of course, that they were more numerous and have subsequently been displaced by the more recent types in which the diploid generation plays a more significant role.

(ii) *Ectocarpus siliculosus* in its English and Mediterranean forms would both begin and end the series, and this hardly seems conceivable. This, however, could not be regarded as a fundamental objection because it might equally well be argued that the species forms an excellent example of how the process of interpolating the diploid generation took place.

(iii) The frequency of parthenogenesis in the Ectocarpales suggests decadence of sexuality rather than the existence of a primitive condition, but it could also be argued that there is a decadence of sexuality in the Laminariales and Fucales.

B. Plants with only a diploid generation, e.g. Fucales, are the most primitive, and the haploid generation has been interpolated subsequently. If this interpretation is correct the only obvious source of origin for the group would be from the Siphonales because a flagellate ancestry would be most unlikely under such circumstances. The evidence that might be adduced in support of this hypothesis is tabulated below:

(i) All the Fucaceae are diploid, and these form a large proportion of the Phaeophyceae and also have an extremely wide distribution.

(ii) In the Laminariales the diploid phase is dominant.

(iii) The haploid phase is frequently omitted in *Dictyota* (cf. p. 165) and also in *Cutleria*.

(iv) The majority of the macroscopic filamentous forms are diploid, the small ectocarpoid filaments forming the haploid generation.

One important objection to this view is the concomitant requirement that the early Phaeophyceae must have started life with a highly complex structure, e.g. *Fucus*, though of course some such structure can be found .in the Siphonales. It must also be remembered that the interpolation of a diploid generation into an original haploid phase may have produced plants that were more successful and which subsequently eliminated their parents in the struggle for existence.

C. The original ancestors were filamentous with equal haploid and diploid generations, or perhaps with generations that were slightly unequal, but that both retained the power of producing a ciliated zooid which could develop without fusion, e.g. *Nemoderma* (fig. 163).

A further consideration of this problem must now be deferred until the other two groups have been surveyed because a final conclusion must incorporate phylogenetic considerations.

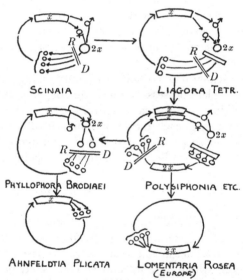

Fig. 164. Types of life cycle in the Rhodophyceae and their possible interrelationships. *RD*=position of reduction division in life cycle.

Fig. 164 is a summary of the principal life cycles that are to be found among the Rhodophyceae. According to Svedelius (1931) the primitive cycle is represented by *Scinaia*, *Nemalion* and *Batrachospermum* where there is only a haploid generation. Some postponement of meiosis is seen in *Liagora tetrasporifera*, but the maximum delay is reached in *Polysiphonia* and most other Rhodophyceae where there are two equal generations, the sporophyte reproducing by means of tetraspores, two of which give rise to male plants and two to female. Subsequent developments, which must be interpreted as retrogressive, can be seen in *Phyllophora membranifolia*, where the tetraspores are grouped into nemathecia on the diploid plant; in *P. Brodiaei*, where the diploid phase has disappeared and the nemathecia can be regarded as growing parasitically in the haploid thallus; and finally in *Ahnfeldtia*, where meiosis no longer takes place and instead the nemathecia contain monospores. *Lomentaria rosea* in European waters is another example of a reduced life cycle, because in this species the gametophyte generation is wholly suppressed, whereas in the other examples it is the sporophyte generation that has been reduced.

In his studies on the Rhodophyceae Svedelius coined a number of terms which have subsequently come into common usage:

Haplobiont. A sexual plant with only one kind of individual or biont, dioecious plants being regarded as representing one kind of individual.

Haplont. A sexual haploid plant with only the zygote diploid.

Diplobiont. A plant possessing alternation of generations and two kinds of individuals, and usually with a much greater number of meiotic divisions since each tetrad of spores involves meiosis. If *Fucus* is regarded as possessing sporangia and reduced gametophytes it will belong to this group rather than being treated as a diploid haplobiont.

Diplont. A sexual diploid plant in which only the gametes are haploid (e.g. *Codium*).

The terms "haplo-" and "diplobiont" do not necessarily coincide with the cytological generations, e.g. *Codium*, and there has been further confusion from the inaccurate usage of these terms by later authors, some of whom have introduced completely new interpretations of the words. In the Rhodophyceae the morphological changes that would be involved make it highly improbable that the diplobionts were primitive to the haplobionts.

Fig. 165 shows a series of typical life cycles that have been found in the principal members of the Chlorophyceae, and here again it will be seen that three principal types can be distinguished:

(1) A multicellular haploid generation in which the diploid phase is present in the unicellular state (e.g. *Ulothrix*).

(2) An alternation between multicellular diploid and haploid generations, both of which are usually morphologically identical.

The only definite exception to this morphological equality at present is seen in *Halicystis ovalis* where *Derbesia marina* forms the diploid generation, although it is possible that a similar state of affairs may exist in *Urospora*.

(3) A multicellular diploid generation in which the gametophyte is reduced to the unicellular state, e.g. *Codium*.

(1 *a*) A persistent unicellular haploid state alternating with a persistent or short-lived unicellular diploid state. This can be regarded as a morphological modification of (1) above or vice versa.

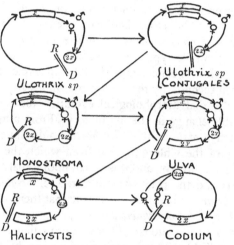

Fig. 165. Types of life cycle in the Chlorophyceae and their possible inter-relationships. RD = position of reduction division in the life cycle.

A study of these life cycles immediately indicates that as a series they can be read in either direction, from 1→2→3 or from 3→2→1. On morphological grounds, however, it is more satisfactory to accept the view that the primitive cycle is that in which the haploid generation is dominant, and that the sporophyte has been subsequently intercalated, presumably by a delay in the occurrence of meiosis as in the Rhodophyceae. Therefore in at least two of the groups it would seem as if the course of events during their evolutionary history has been much the same. In the primitive state the haploid filaments would perhaps be monoecious so that the first development would concern the appearance of the dioecious condition, e.g. *Ulothrix* sp. and the Conjugales.

Monostroma represents another intermediate condition in which the enlargement of the zygote can be regarded as an incipient delay before the reduction division takes place.

SUMMARY AND CONCLUSIONS

In fig. 166 are set out some simplified diagrams of the life cycles of the principal algal types to be found in all three groups. They have all been drawn up on the same principle so that comparisons will be rendered easier. On the hypothesis that the Chlorophyceae are probably the original ancestors of most of the algal groups, the types of life cycle to be found there have been made the basis of the other diagrams. *Chlamydomonas*, *Ulothrix* and *Coleochaete* can all be regarded as simple types in so far as their life cycles are concerned, although it is conceivable that the life cycle of *Coleochaete* may have been secondarily reduced to the wholly haploid stage, or, more probably, that morphological evolution took place without any comparable change in the life history. From a morphological and reproductive standpoint *Coleochaete* would appear to be the only member of the Chlorophyceae from which the Eu-florideae might be evolved directly, and it is worth noting that the life cycles of *Coleochaete* and the primitive Eu-florideae, *Scinaia*, *Nemalion*, *Batrachospermum*, are identical. It is true that there are differences in structure between *Coleochaete* and the primitive red algae, but so long as there is a complete lack of any intermediate stages it is not necessarily justifiable to abandon such an origin because there is an equal lack of intermediate stages for any other source of the Rhodophyceae (cf. p. 256). It would seem, therefore, that a study of the life cycles of the Chlorophyceae and Rhodophyceae can lead one to two conclusions:

(1) Their phylogenetic history follows parallel lines whereby they commence with a wholly haploid generation and the diploid generation is subsequently interpolated through a delay in the occurrence of meiosis. Svedelius (1931) has suggested that the delay in meiosis came about gradually, but cytologically it is perhaps easier to imagine one or more sudden delays resulting in two morphologically similar generations, one of which, the diploid, subsequently may have undergone modification.

(2) There are grounds for believing that some of the filamentous

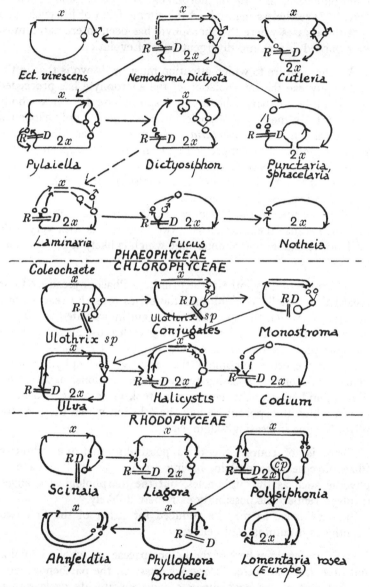

Fig. 166. The principal types of life cycle in the Phaeophyceae, Chlorophyceae and Rhodophyceae and their possible inter-relationships.

Chlorophyceae, in spite of morphological differentiation (hetero-trichy), nevertheless maintained the simple form of life cycle, and that those cases where the sporophyte has been interpolated must be regarded as forming divergent lines of evolution.

Another feature to which attention must be drawn is that in the Chlorophyceae the interpolation of the sporophyte has proceeded considerably further, whereby the sporophyte becomes wholly dominant. In the Rhodophyceae, however, this has only happened in one case, namely *Lomentaria rosea*, and even here the gameto-phyte has only been suppressed abnormally in European waters. In the red algae there is a reduction series instead, and this leads back to wholly haploid plants, e.g. *Ahnfeldtia*, in which the con-dition has been produced secondarily.

When we turn to the Phaeophyceae the problem is much more difficult because there are at least two alternatives with very little evidence to enable one to determine which is likely to be the more correct:

(1) On the first hypothesis the primitive Phaeophyceae are to be regarded as wholly haploid, and the series must be read in one direction in which the sporophyte is again interpolated through a delay in meiosis, the series terminating with those algae in which the sporophyte generation is wholly dominant, e.g. *Fucus.*

(2) On the other hypothesis the primitive Phaeophyceae were filamentous forms possessing two equal generations, haploid and diploid, and subsequent development took place along two lines, one in which the sporophyte and the other in which the gameto-phyte became increasingly dominant.

There is, of course, the third possibility that the primitive Phaeophyceae were diploid, having arisen from diploid Chloro-phyceae such as the Siphonales, but the morphological changes involved render this possibility extremely unlikely.

Such evidence as may be adduced for either of the first two hypotheses is summarized below:

(a) Very few members of the Ectocarpaceae are wholly haploid, and in at least one case, *Ectocarpus virescens*, the parthenogenetic development of the eggs suggests a degenerate life cycle rather than a primitive one.

(*b*) Some of the primitive forms, e.g. *Lithoderma* and *Nemoderma*, have two equal generations in the life cycle and a similar state of affairs is also found among other brown algae, e.g. *Dictyota*, *Zanardinia*.

(*c*) The ultimate decision must obviously be largely determined by the condition of affairs found in the sources from which the Phaeophyceae arose. On general grounds it is to be supposed that the Phaeophyceae all arose from one common ancestor, but it must not be forgotten that the group may have had a polyphyletic origin, although at present there is hardly any evidence in support of such a view. Two possible sources of origin for the Phaeophyceae have been suggested in the past. One is that they arose, as did the Chlorophyceae, from a flagellate ancestry with intermediate forms such as *Phaeococcus* and *Phaeothamnion* (cf. p. 123). On the basis of their pigments the Chrysophyceae (cf. p. 122) show a close resemblance to the more primitive Phaeophyceae and this is not without significance. If this theory is correct, one must almost certainly consider that the primitive species, as in the primitive Chlorophyceae, were wholly haploid and that the diploid state has been interpolated subsequently. The other hypothesis is that they arose from some member of the Chlorophyceae, probably among the Chaetophoraceae. This latter group is characterized by heterotrichy, a feature which is possessed by some of the primitive Phaeophyceae, whilst another point in favour of this view is the lack of any satisfactory existing series between the few known phaeophycean-like flagellates and the primitive filamentous Ectocarpales. If we accept an origin of the Phaeophyceae from the Chlorophyceae, two possible sources may be suggested:

(*a*) From a member of the Chaetophoraceae which possessed the heterotrichous habit and two morphologically similar or nearly similar generations.

(*b*) From a member of the Siphonocladiales which had a life cycle with two equal generations, such as is now shown by *Chaetomorpha* or *Cladophora Suhriana*.

It is tempting to consider whether the Phaeophyceae have not been derived from a form such as *Trentepohlia*, and it is much to be regretted that at present the life cycle of *Trentepohlia*, so far as cytological details are concerned, is wholly unknown. Until we possess a more extensive knowledge of the cytological life cycles among the Chaetophoraceae it would appear futile to speculate further on the origin of the Phaeophyceae, and all that can usefully be done in this chapter is to point out the various possibilities. One further point remains to be added. In the present volume it has been suggested that the similarity in life cycles and phylogenetic histories leads one to the hypothesis that the three groups of algae are perhaps intimately related. At the same time an attempt has been made to indicate that there are other workers who believe that all three groups have had independent origins from different sources, and that the various types of life cycle have evolved independently. At present the decision between these two courses would seem to be largely a matter of opinion.

*EVOLUTION

Rhodophyceae

It has already been suggested above that the primitive Rhodophyceae, in particular the Eu-florideae, may have arisen from a member of the Chaetophoraceae such as *Coleochaete*. It is only proper, however, to emphasize that this is purely one viewpoint, and that there are other workers who have sought for an origin of the group from among the unicellular organisms, but unfortunately there are very few members of the Protista which can be regarded as possible sources for the red algae apart from *Porphyridium cruentum*. An alternative hypothesis is that which considers the Rhodophyceae to have been evolved from the Cyanophyceae, the principal argument in support of this view being the resemblance between the colouring pigments, primarily in colour because it has recently been shown that the pigments are not identical chemically. The principal objection to this theory is the absence of any form of sexual reproduction among the Cyanophyceae, a feature which renders the presence of highly specialized sex organs in even the most primitive Rhodophyceae difficult to explain. Apart from these theories, however, there is also the possibility that the Eu-florideae have been evolved from the Proto-florideae, in which case the origin of the latter group becomes of importance. Two possible lines of evolution can be suggested, but it does not appear feasible to discriminate in favour of either one:

(1) A. *Aphanocapsa→Porphyridium→Porphyra.*
    B. *Prasiola→Bangia→*Eu-florideae.

Apart from the difference of pigment there is a striking resemblance in morphological structure and reproduction between *Bangia* and *Prasiola.*

(2) Cyanophyceae → *Porphyridium* → *Porphyra* → *Bangia* → Eu-florideae.

In contrast to this there are those who postulate independent origins for the Proto- and Eu-florideae on account of the considerable differences in structure and reproduction between the two divisions, but at present it would seem impossible to do more than point out the different theories that have been put forward. The later evolutionary changes in the Eu-florideae have already been mentioned in the introduction to Chapter VIII and also in the earlier part of the present chapter.

PHAEOPHYCEAE

Most workers would probably agree that the primitive members of this group are to be found among those members of the Ectocarpales which either possess a single haploid generation or else two morphologically identical generations. It now remains to indicate how subsequent evolution may have taken place, but only a broad outline can be suggested because individual workers have frequently produced modifications in the lesser details of the evolutionary sequences. It has been stated that the unilocular sporangia of the Ectocarpales, in which meiosis occurs, are morphologically equivalent to the tetrasporangia of the Dictyotales. In this case the plurilocular sporangia which give rise to the iso- or anisogametes are morphologically equivalent to the gametangia of the Cutleriales and Dictyotales. One of the outstanding problems is the origin of the Laminariales and Fucales, and in order to account for these it would seem necessary to postulate at least two different lines of evolution, though there were probably even more. As an example of the simpler type of sequence that has been suggested the following may be quoted from Svedelius:

Phaeosporeae→Cutleriales→Dictyosiphonales→
Dictyotales→Laminariales→Fucales.

Kylin (1933), on the other hand, whilst agreeing with an origin for the Dictyotales from the Phaeosporeae suggests that the Fucales have not arisen from that source. It is possible to imagine a line of evolution, not only on morphological, e.g. the cable type of construction, but also on reproductive criteria, commencing from *Ectocarpus→Castagnea→Chordaria→Chorda→Laminaria*, whilst an alternative source for the Laminariales could also be found in parenchymatous genera such as *Dictyosiphon* or *Punctaria*. It is extremely tempting to consider whether the Fucales may not have been evolved from the Laminariales because of the existence of forms such as *Durvillea*, and if the oogonia and antheridia of the Fucales are regarded as modified unilocular sporangia, e.g. micro- and macrosporangia, then this becomes a possibility. On the other hand, the oogonia and antheridia might be regarded as more closely allied to the tetrasporangia of the Dictyotales which must then be regarded as a possible ancestral source.

TABLE III

Ectocarpales	Dictyotales	Laminariales	(*Macrocystis*)	Fucales
Plurilocular sporangia	≡ Oogonia antheridia	≡ Oogonia antheridia	≡ Oogonia antheridia	≡ Oogonia antheridia
Unilocular sporangia	≡ Tetrasporangia	≡ Unilocular sporangia (homosporous)	≡ Unilocular sporangia (heterosporous)	≡ Micro- and megasporangia (heterosporous?)

From the above table it would seem that the evolution of the Fucales is associated with the development of heterospory in much the same way as the evolution of the seed habit is often said to be associated with the development of heterospory. The origin of such a habit forms a very distinct problem because there is very little evidence for such a development among the other phaeophycean groups. In the Laminariales heterospory but not heterangy is recorded for *Macrocystis*, and if it has arisen once it may have been present in some of the ancestral Laminariales, forms from which perhaps the Fucales arose. It is also possible that we are pursuing a false scent in trying to establish heterospory as a feature of the Fucales, and that if the actual spores are considered, e.g. the products of the first two divisions in the micro- and megasporangia, it would be found that they are really homosporous, in much the same way as Thomson has suggested that the so-called micro- and megaspores of the angiosperms are really homosporous. If this were found to be true, then the problem of the origin of heterospory in the Fucales would be disposed of and the evolutionary problem much simplified. It is also possible that the explanation of heterospory and heterangy in the Fucales is to be found in the retention on the parent thallus of heterothallic gametophytes, the stimulus provided by their presence being responsible for the modification of the original morphologically identical unilocular sporangia.

One or two authors have recently suggested that an origin for the Fucales should be sought for among the Mesogloiaceae and Encoeliaceae (*Colpomenia*), but the evidence produced cannot be regarded as wholly convincing. The most recent account by Delf (1939) considers this problem in some detail. In adult plants of *Fucus* there is an apical growing cell which is now known to arise as follows. In the sporeling a group of apical hairs is formed at the growing point, each hair possessing basal (trichothallic) growth as in the Ectocarpales. These hairs die off and the lowest cell of one hair gives rise to the four-sided apical cell of the adult thallus. A similar behaviour of the apical hairs is to be seen in *Acrothrix* (Mesogloiaceae). New growth from wounded tissue in *Fucus* also develops a new apical cell from such trichothallic hairs, whilst the development of the cryptostomata and conceptacles also appears to be analogous. It is further suggested that in gross structure, e.g. primary and secondary medullary filaments and the assimilatory tissue, the thallus of *Fucus* shows considerable resemblance to that of *Eudesme* as illustrated in fig. 95. Difficulties associated with this interpretation must be concerned with the differences in size of the thalli and also the presence of heterospory and heterangy. The gametophytes of the Mesogloiaceae reproduce by means of plurilocular sporangia which do not exhibit either anisogamy or heterangy nor do the sporophytic plants exhibit either heterospory or heterangy. Recent work, however, has shown that the gametophyte of *Colpomenia sinuosa* bears organs that must be regarded as relatively simple antheridia and oogonia, so that there is here an example of heterangy associated with anisogamy (cf. p. 154). Whether the Fucales are derived from the Laminariales or

Dictyotales, it is obvious that their ancestry is not to be found in the present living forms of either group. It would therefore seem best for the time being to derive the Fucales from the ancestral groups of either the Laminariales or Dictyotales, recognizing that there is a definite bridge in both cases, the gap being least perhaps between the Fucales and Laminariales. The common race cycle found in the Phaeophyceae with its irregular alternation of generations must have evolved several times, the course of evolution probably being determined by the morphological changes, e.g. corticated type, reduced ectocarpoid type, reduced cable type.

It has been pointed out that the Phaeophyceae can be divided into two great groups, the Isogeneratae and Heterogeneratae, and the latest schemes of evolution take these into consideration. In both Iso- and Heterogeneratae there is a gradual transition from isogamy to anisogamy, and on these grounds one can perhaps postulate at least two major lines of evolution. The schema below is an example of what can be obtained employing this line of approach, which is probably more satisfactory than one that is purely morpho-logical:

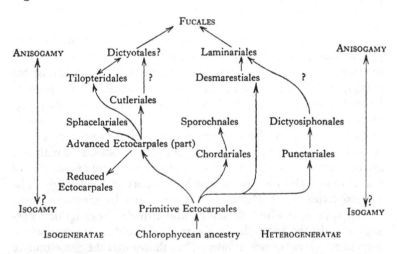

The names of the orders given above do not imply that the present living representatives formed the stages in evolution, but that types more or less similar to them existed in the evolutionary sequence.

CHLOROPHYCEAE

Among the primitive forms a schema such as the one below (modified after Senn) will give an indication of the primary sequence of events:

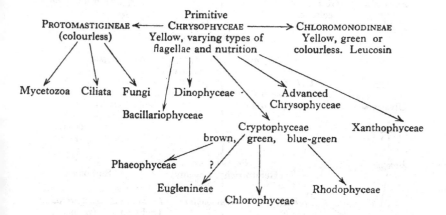

The origin of the Phaeophyceae and Rhodophyceae has already been discussed in some detail and hence will not be considered further.

With this scheme as a basis it is now possible to consider evolution among the Chlorophyceae, and here there are two starting points. One hypothesis commences with the Chlamydomonadaceae and branches out with a number of lines of evolution, whereas in the other the Palmellaceae are regarded as the source of the group. The first of these two theories is perhaps the more satisfactory. The origin of the Siphonales offers a similar kind of problem to that of the Fucales because there are very few satisfactory intermediate forms. As a result they are illustrated in the schema as having been evolved either directly from an unicellular organism or else from the Siphonocladiales, the latter perhaps forming the more attractive hypothesis. It has recently been emphasized that the nearest affinities, even though distant, of the Siphonocladiales, are with the Siphonales via *Valonia* and *Halicystis* on the one hand, and with *Ulothrix* via *Chaetomorpha* on the other.

SCHEME A

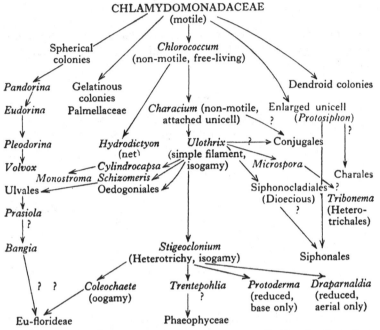

The names of the genera do not necessarily imply that they
formed the actual intermediate stages, but merely that forms like
them existed in the evolutionary sequence.

SCHEME B

The Chaetophorales are represented as derived either from the
Ulotrichales (A) or from the Palmellaceae (B), but Vischer has

suggested that they are derived from the Volvocales on the basis of a resemblance between them and the simpler forms, e.g. *Gongrosira*. It is, however, probably more correct to interpret these simpler forms as reduced rather than primitive.

From the above two schemas it will be seen that there are a number of definite morphological tendencies, and it has already been pointed out that these various lines of morphological development are repeated in the different algal groups. Table IV from Fritsch (1935) provides examples of parallelism in evolution among the simpler types of algae.

Apart from these examples of evolution among the simpler algae it is also found that other evolutionary tendencies can be observed among the more advanced types of algae. Such tendencies are illustrated in Table V.

The concept of the heterotrichous habit was first advanced by Fritsch in 1929, and it has become increasingly evident that an understanding of this habit is of fundamental importance in considering any phylogenetic or evolutionary problem among the algae. In the primitive state both the prostrate and erect systems must be present, but during the course of evolution one of these has frequently become reduced or lost, e.g. in *Endoderma* (Chlorophyceae), *Streblonema* (Phaeophyceae) and *Melobesia* (Rhodophyceae) only the prostrate system remains. In contrast to this the thallus in *Draparnaldia* and most of the Florideae represents the erect system, the prostrate system having been reduced or lost. Another fact in connexion with this phenomenon, which needs to be re-emphasized, is that the most advanced Chlorophyceae exhibit the heterotrichous habit in its primitive state, whilst this condition is only found fully developed among the simpler Rhodophyceae and Phaeophyceae since in the more evolved types one or other of the systems is reduced. The possible implications of this observation are immediately obvious.

Finally, a word may be said about the time when the different groups first made their appearance. Most authors would consider that the Cyanophyceae and Chlorophyceae are the most primitive and therefore appeared first. If, however, the Rhodophyceae and Phaeophyceae have a flagellate origin then all four groups may be of almost the same antiquity. There are some workers who believe that the Cyanophyceae are the most primitive group and that they

TABLE IV. *Parallelism in evolution of the simpler type of algal construction*

Type of construction	Chlorophyceae *Chlamydomonas	Xanthophyceae Heterochloris	Chrysophyceae Chromulina	Cryptophyceae Cryptomonas	Dinophyceae Gymnodinium
(1) Motile naked green unicell	*Chlamydomonas	Heterochloris	Chromulina	Cryptomonas	Gymnodinium
(2) Encapsuled unicell	*Phacotus	—	Chrysococcus	—	—
(3) Motile colony	*Pandorina, *Volvox	—	Symura	—	Polykrikos
(4) Dendroid colony	Prasinocladus	Mischococcus	Chrysodendron	—	—
(5) Palmelloid colony	*Tetraspora	Chlorosaccus	Chrysocapsa	Phaeococcus	Gloeodinium
(6) Coccoid forms	*Chlorococcum, *Chlorella	Chlorobotrys	Chrysosphaera	Tetragonidium	Cystodinium
(7) Simple filament	*Ulothrix, *Urospora	*Tribonema	Nematochrysis	—	Dinothrix
(8) Branched filament	*Cladophora	—	—	—	—
(9) Heterotrichous filament	*Stigeoclonium	—	—	—	Dinoclonium
(10) Siphonaceous filament	*Codium, *Vaucheria	*Botrydium	—	—	—

* It is recommended that elementary students should only remember these types and the fact that there are examples in other groups.

gave rise later to the Florideae. These, it is then supposed on fossil evidence together with the similarity in pigmentation, were followed by the Phaeophyceae and Chrysophyceae, the Chlorophyceae being the last group to appear. It must be pointed out, however, that the absence of fossil remains does not necessarily mean that a group was absent at any given period: many of the Chlorophyceae are delicate forms and would not be preserved so readily as the tougher fronds of the brown and red algae. The oldest fossil Chlorophyceae to be recognized belong to the Siphonales and Siphonocladiales and they show a high degree of differentiation which suggests that, as a group, they were evolved at a

TABLE V. *Parallelism in evolution among the advanced types of algae*

Type of construction	Chlorophyceae	Phaeophyceae	Rhodophyceae
(1) Heterotrichous filament	*Stigeoclonium*	*Ectocarpus*	*Batrachospermum*
(2) Discoid type	*Protoderma*	Ascocyclus	Erythrocladia
(3) Crusts or cushions	Pseudopringsheimia	Ralfsia	Hildenbrandtia
(4) Elaborated erect type	*Draparnaldia*	Desmarestia	Callithamnion
(5) Compact pseudoparenchymatous type, *uniaxial*	Dasycladus *Chara*	Mesogloia	*Ceramium*
(6) Ditto, *multiaxial*	*Codium*	Castagnea	Nemalion
(7) Foliose parenchymatous type	*Ulva*	Punctaria	*Porphyra*
(8) Tubular parenchymatous type	*Enteromorpha*	Asperococcus	Halosaccion

* It is recommended that elementary students should only remember these types together with the fact that the others do exist.

very early stage. With very little evidence to support the view, certain workers consider that these siphonaceous fossils represent the primitive green algae, the remainder, principally fresh-water forms, developing much later. Whilst it is doubtful if many algologists would subscribe to this interpretation, it is mentioned here because it is felt that any suggestions, however likely or unlikely, open up fresh fields of thought and investigation.

On the more orthodox interpretation it would seem as if the Cyanophyceae and unicellular Chlorophyceae were the most primitive algae. Some of the Florideae, especially the Protoflorideae, may have appeared quite early, whilst the Eu-florideae perhaps developed somewhat later at the same time as the

Phaeophyceae. Whatever the sequence of events, it is quite clear from the structure of the earliest fossils that considerable evolution had taken place long before their time.

FOSSIL FORMS

In this section it is merely proposed to give an outline of the different fossil forms that have been ascribed to the various groups, but it is not intended to provide a detailed description in every case so long as the types of structure represented among these fossil algae have been adequately portrayed. It must be realized that many of the early forms that have been ascribed to the algae are relatively unknown because of the poor preservation, and further examination of new specimens may mean that they will have to be removed from the algae. For this reason it must be emphasized that there are a number of doubtful forms from the lowest strata which can only be tentatively assigned to the algae.

CYANOPHYCEAE

Among the unicellular forms a fossil which has been related to the Chroococcaceae is recorded from the Ordovician. It is called *Gloeocapsomorpha* and is a colonial form with cells that were apparently enclosed in a jelly, and whilst it may have affinities with living colonial forms it is usually placed in a group called the Protophyceae. Another plant of Middle Cambrian age, *Marpolia spissa* (fig. 167), which seems to have affinities with the modern *Schizothrix*, is also best relegated to the Protophyceae. *Marpolia* was represented by branched filaments which were probably composed of a trichome enclosed within a gelatinous or cartilaginous sheath.

Spongiostromata (Precambrian to present day).

Much doubt has been thrown upon the authenticity of this group, some writers regarding them as structures which originated as diffusion rings ("liesegang" phenomena)

Fig. 167. *Marpolia spissa* (× 49·5). (After Walton.)

in colloidal materials or perhaps in calcareous muds. In the original description Walcott suggested an affinity to the Cyanophyceae, but as later workers could only distinguish a purely mineral structure they suggested the idea of diffusion phenomena. Discoveries of very comparable algal concretions and laminations in the Bahamas, however, have made it extremely probable that these structures had an algal origin. Some examples of these types are shown in fig. 168. On the basis of Black's discoveries (1933) it may be suggested that these structures were not necessarily formed by deposition but that the algae collected and bound the sediment.

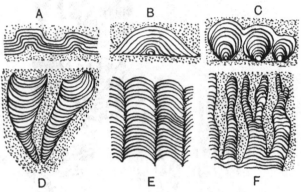

Fig. 168. Stromatolithi. A, *Weedia*. B, *Collenia*. C, D, *Cryptozoon*. E, *Archaeozoon*. F, *Gymnosolen*. (After Hirmer.)

Porostromata (e.g. *Girvanella*, *Sphaerocodium*).

These forms, which are most abundant in the Carboniferous, have a recognizable microscopical structure, the threads often being arranged in a radiating fashion: they were probably formed in much the same way as the algal water-biscuits now found in South Australia. These range from tiny particles to thick bun-like forms 20 cm. in diameter, whilst in them are to be found the tube-like remains of living species of *Gloeocapsa* and *Schizothrix*. *Gloeothece* and *Gloeocapsa* are also known to form oolitic granules in the neighbourhood of Salt Lake City. The presence, however, of pebbles, or the existence of a granular structure, does not necessarily involve the presence of algae, and in some cases it is also possible that the algae were merely included through chance. *Pachytheca* is a genus from the Silurian and Devonian which possesses a medulla of intertwining tubes and a cortex composed of stout, septate, branched algal filaments that radiate from the medulla to the periphery. Its affinities are extremely uncertain and it may have been a free-rolling alga of either salt or fresh waters (cf. fig. 169).

There are a few uncertain fossils, very indistinct and not well known, ascribed to the Flagellata and Dinophyceae. Recognizable

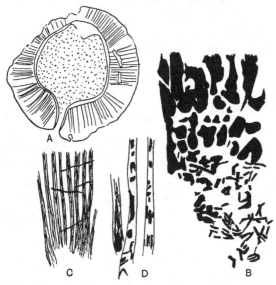

Fig. 169. *Pachytheca*. A, transverse section with natural opening through cortex (× 12). B, algal filaments of medulla and inner cortex (× 240). C, cortex with algal filaments (× 60). D, cortex showing degenerate algal threads in tube (× 150). (After Lang.)

fossil diatoms are known from the Upper Jurassic, and there was a very rich fossil diatom flora in the Tertiary, all the specimens found being closely related to existing families and genera.

Codiaceae

Boueina (cf. fig. 170) is an unbranched form from the Lower Cretaceous, whilst *Palaeoporella* (fig. 170), which is composed of hollow cylinders or funnel-shaped bodies with slender forked branches, the whole being two to fourteen millimetres long, comes from the Lower Silurian. *Dimorphosiphon*, from the Ordovician, is generally regarded as the oldest known member of the Codiaceae and has been tentatively related to *Halimeda*. It is about ten millimetres long and is composed of branched tubular cells without any cross walls, the cells being embedded in a calcareous matrix. *Ovulites*, a genus which occurs up to the Eocene, differs considerably from those previously described: the species are little egg or club-like chalk bodies beset with fine pores and with a large opening at what was either the base or apex. It has been suggested that

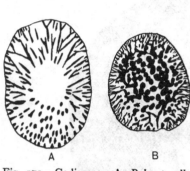

Fig. 170. Codiaceae. A, *Palaeoporella variabilis* (× 12). B, *Boueina Hochstetteri.* (After Hirmer.)

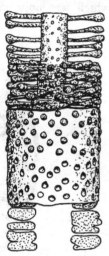

Fig. 171. Dasycladaceae. *Rhabdoporella pachyderma* (× 135). (After Hirmer.)

perhaps they represent siphonaceous plants in which the apical tuft of threads has been lost.

Dasycladaceae

This is the best known group and contains a very large number of the fossil algae. It reached its maximum development and abundance in Carboniferous and Triassic times, and in those days was far more important than its present living representatives. The various forms are all based on a type of construction which can be sufficiently explained by descriptions of a few of the more representative types.

Rhabdoporella (fig. 171) seems to be one of the most primitive genera as it is represented by a purely cylindrical shell that is studded with pores through which the threads passed. It is known from the Ordovician and Silurian.

Cyclocrinus (fig. 172) is a genus which grew to about seven centimetres and looked like a miniature golf ball borne on the end of a stalk. Narrow branches arose at the apex of the stalk and each terminated in a flattened hexagonal head, but as the edges of adjoining heads were fused together to form the outer membrane, which was only weakly calcified, the cell outlines were clearly

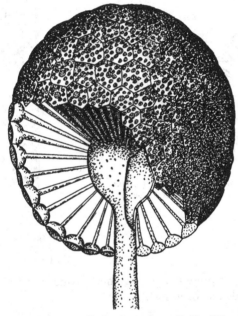

Fig. 172. Dasycladaceae. *Cyclocrinus porosus* (× 8). (After Hirmer.)

visible. Many species are known from the Ordovician and Silurian, all somewhat resembling the living genus *Bornetella*.

Primicorallina (fig. 173), from the Ordovician, had a segmented stem beset with radially arranged branches, each of which branched twice into four branchlets.

The type of structure found in *Diplopora* (fig. 174) was also shown by many other forms from the Middle Triassic. It was a few centimetres long and bush-like in appearance, the main stem, which sometimes had a club-shaped apex, being covered with whorls of branches that arose in groups of four, each bearing secondary branches which terminated in hairs. In the older thalli the outer part of the branch dropped off leaving a scar on the calcareous shell. The sporangia are reported to have been modified branches. *Diplopora* is a widespread genus from the Triassic rocks of the eastern Alps, Germany and Siberia.

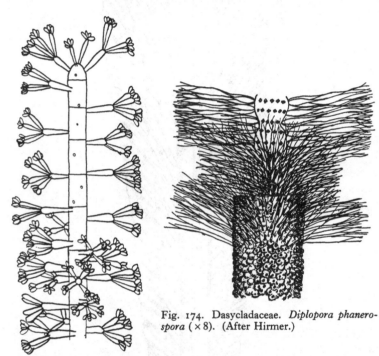

Fig. 174. Dasycladaceae. *Diplopora phanerospora* (× 8). (After Hirmer.)

Fig. 173. Dasycladaceae. *Primicorallina trentonensis* (× 8·25). (After Hirmer.)

Palaeodasycladus (fig. 175), from the Lower Jurassic, bears a resemblance to the living species of *Dasycladus*. Near the base there were only primary branches, whilst higher up secondary and tertiary branches were to be found.

Fossil forms, practically identical with living species of *Cymopolia*, *Neomeris* and *Acicularia* have been found in all the recent strata from the Eocene upwards.

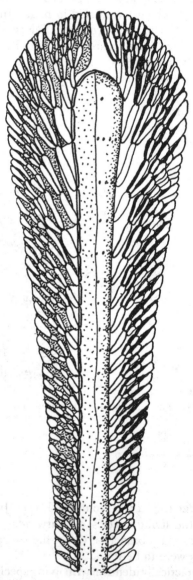

Fig. 175. Dasycladaceae. *Palaeodasycladus mediterraneus* (× 20). (After Hirmer.)

Charophyta

Lagynophora, a genus from the Lower Eocene, can be ascribed to this group, whilst *Palaeonitella* (fig. 176), from the Middle Devonian,

Fig. 176. Charales. *Palaeonitella Cranii* (× 124). (After Hirmer.)

may belong here also although its affinities are not so clear. *Gyrogonites* and *Kosmogyra* are names which have been given to oogonial structures which closely resemble those of *Chara*, and which are very abundant in the Lower Tertiary beds of England and elsewhere.

Phaeophyceae

The principal fossil form ascribed to this group, *Nematophyton*, has now been removed to a new group, the Nematophytales (see p. 274).

Rhodophyceae

The Melobesiae are represented from the Cretaceous upwards by species of *Archaeolithothamnion, Dermatolithon, Lithothamnion, Lithophyllum* and *Goniolithon*, some of them only being distinguished with difficulty from living forms. The Corallinaceae are also represented in the Cainozoic by extinct members of present living genera. There are a large number of forms assigned to an extinct family, the Solenoporaceae, which existed from the Ordovician up to the Triassic, but neither their structure nor their systematic position has been completely established. They formed nodules from the size of peas up to several centimetres in diameter in which the cells were arranged like those of a *Lithothamnion* although the cross walls were not well marked.

NEMATOPHYCEAE: Nematophytales. Figs. 177, 178.

Two genera are now grouped in this assemblage which has recently been established by Lang (1937), and although he regards these forms as land plants, nevertheless they have so many features

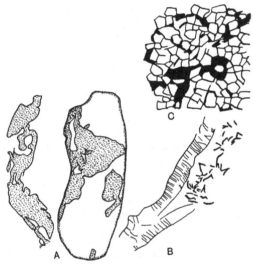

Fig. 177. Nematophytales. *Nematothallus.* A, specimens on rock (× ⅔). B, large and small tubes, the former with fine annular thickenings (× 150). C, cuticle (× 150). (After Lang.)

in common with the algae that it is felt proper to include them here. It is perhaps almost too speculative to suggest that they represent Church's transmigrant form, but it would appear that they must either be regarded as highly developed algae which adopted a land habitat, or else as the most primitive of all true land plants. The two genera agree closely in their morphological structure, and although they are both frequently found associated with each other in the Devonian rocks the two structures have not yet been found in organic connexion. In spite of this it is very probable that the leafy *Nematothallus* was the photosynthetic lamina of the stem-like *Nematophyton* and may also have functioned as the reproductive organ. In the lowest strata the plants are to be found associated with remains of marine animals, thus suggesting their power to grow under marine or brackish conditions, whilst in the higher strata they occur in beds, which are regarded as fresh-water or continental, where they are associated with plants that were undoubtedly terrestrial. The presence of spores in *Nematothallus* is regarded as rendering it unlikely that they were algal in nature, but the spores may be comparable to the hard-walled cysts such as are to be found in *Acetabularia*.

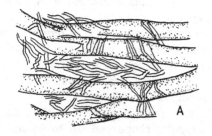

Fig. 178. Nematophytales. *Nematophyton*. A, longitudinal section (× 120). B, transverse section (× 120). (After Seward.)

The genus *Nematophyton* is found in the Silurian and Devonian rocks where it was first described under the name of *Prototaxites* and referred to the Taxaceae, but subsequently it was accepted as an alga and renamed *Nematophyton* or *Nematophycus*. Later the name *Prototaxites* was revived and it was placed in the Phaeophyceae, whilst Kräusel (1936) recently stated that it must have had the appearance of a *Lessonia* (cf. p. 180) and also that it existed in aquatic habitats which may have been marine, brackish or fresh. The valid name is therefore *Prototaxites*, but as this tends to convey a false impression of the plant's affinities it would seem more satisfactory to retain the better known name of *Nematophyton*. The largest specimen is a stem up to two feet in diameter, but whatever the size of the stem they are usually composed of two kinds of tubes, large and small. The large tubes have no cross partitions, but in some species they are interrupted in places by areas, regarded as medullary rays or spots by some authors, which are wholly occupied by small tubes that in other parts of the thallus simply take a sinuous course between the large tubes. The wide tubes, in the latest specimens described by Lang (1937), show no markings indicative of definite thickening, though striations have been seen in specimens from other localities. Around the outside of the central tissue there is a cortex, or outer region, composed of the same tubes where they bend outwards towards the periphery and eventually stand at right angles to the surface. The outermost zone of all is apparently structureless and may well have been a mucilaginous layer during life.

Nematothallus is a genus composed of thin, flat, expanded incrustations of irregular shape and up to 6½ cm. long by 1 cm. broad, and also constructed of the wide and narrow tubes. The thallus is surrounded by a cuticular layer that exhibits a pseudocellular pattern, and which includes within the cuticle and among the peripheral tubes firm-walled spores of various sizes; in *N. pseudovasculosa* the spores were definitely cuticularized and so the suggestion was made that these were land plants or parts of a land plant. The wide tubes, which have thin pale brown walls, are translucent in appearance and exhibit distinct characteristic annular thickenings. The cuticle, which is apparently readily detached, possesses distinct cell outlines that were probably made by the ends of the wide tubes from the ordinary tissue where they became fused together at the periphery, as in the living genera *Udotea* and *Halimeda*. Another species *Nematothallus radiata* is more imperfectly known.

From the structure described above it can be seen that the members of this group are strongly reminiscent of the Laminariales and Fucales, and it is tempting to suppose that they represent land migrants from one of these groups. Problems that have to be solved are: (1) The cuticularized spores; whilst no such spores with hard outer walls are known from the brown algae they are recorded from the Chlorophyceae, e.g. *Acetabularia*. However, the suggestion that the spores may have developed in tetrads adds a further complication, at any rate so far as an algal ancestry is concerned, because the Dictyotales and tetrasporic Rhodophyceae do not show the state of differentiation found in these fossil plants. (2) The presence of a deciduous cuticle. In this connexion one or two Laminariales are known to shed cuticles during reproduction, and the present author has found a deciduous cuticle on some preserved plants of *Hormosira*, a member of the Fucales. It may be suggested that the plants perhaps had the appearance of a *Lessonia* or even of a *Durvillea*, and a stem diameter of two to three feet does not preclude them from being algal in character because several of the large Pacific forms may have stipes of almost this size (cf. p. 180). It has also been suggested that these forms are related to the Codiaceae, especially *Udotea*, and in certain respects it is true that they have the structure of a siphonaceous plant. Here again there are several problems that need to be answered: (*a*) the presence of two sizes of tubes; (*b*) the presence of a cuticle; (*c*) the presence of cuticularized spores; (*d*) the large size of stem.

The answer to the last problem has already been suggested (see above) but cuticles in the Codiaceae have not been recorded, although the present author has been able to detect a structure something like a cuticle in *Halimeda*; nor have any species been reported that possess two distinct sizes of tubes, although gradations in size occur in both *Udotea* and *Halimeda*. In this connexion it may be of interest to refer to Tilden's unsupported suggestion that the land plants arose from forms such as *Codium* and *Caulerpa*. It must be admitted that there are no living members of the Codiaceae with stems that approach anywhere near the size of those of *Nematophyton*. This, however, is not an insuperable objection as the Nematophytales may bear the same relation to the living Codiaceae that the fossil Lepidodendrons bear to the living Lycopodiales. For the present, however, the problem must be left in the hope that further evidence will accumulate.

REFERENCES

Spongiostromata. BLACK, M. (1933). *Philos. Trans.* B, **222**, 165.

Evolution. DELF, M. (1939). *New Phytol.* **38**, 224.

Evolution. FRITSCH, F. E. (1935). *Structure and Reproduction of the Algae*, vol. 1, p. 12. Camb. Univ. Press.

Heterotrichy. FRITSCH, F. E. (1939). *Bot. Notiser*, p. 125.

278   REPRODUCTION, EVOLUTION, ETC.

Reproduction. KNIGHT, M. (1931). *Beih. bot. Zbl.* **48**, 15.
Nematophytales. KRÄUSEL, R. and WEYLAND, H. (1934). *Palaeonto-graphica*, **79**, Abt. B, p. 131.
Nematophytales. KRÄUSEL, R. (1936). *Ber. dtsch. bot. Ges.* **54**, 379.
Reproduction. KUNIEDA, H. and SUTO, S. (1938). *Bot. Mag., Tokyo,* **52**, 539.
Evolution. KYLIN, H. (1933). *Lunds Univ. Årsskr.* N.F. Avd. 2, **29**.
Nematophytales. LANG, W. H. (1937). *Philos. Trans.* B, **227**, 245.
Dasycladaceae. PIA, J. (1920). *Abh. zool.-bot. Ges. Wien*, **11**.
Fossils PIA, J. (1927). In Hirmer's *Handb. Palaeobot.* München and Berlin.
Fossils, General. SEWARD, A. C. (1931). *Plant Life through the Ages.* Cambridge.
Evolution. SMITH, G. M. (1933). *Fresh Water Algae of the United States*, p. 4. New York.
Reproduction. SMITH, G. M. (1938). *Bot. Rev.* **4**, 132.
Reproduction. SVEDELIUS, N. (1927). *Bot. Gaz.* **83**, 362.
Reproduction. SVEDELIUS, N. (1931). *Beih. bot. Zbl.* **48**, 38.
Evolution. TILDEN, J. (1935). *The Algae and their Life Relations*, p. 24. Univ. Minn. Press.

CHAPTER X

PHYSIOLOGY, SYMBIOSIS, AND SOIL ALGAE

PHYSIOLOGY

It would obviously be impossible to attempt a complete survey in these pages of all that is known concerning the physiology of the algae, especially as many species are very suitable objects for the study of certain branches of physiology and in such cases a voluminous literature has accumulated. *Valonia* has been frequently used in experiments on absorption of solutes because of the large size of the vesicles; the Charales have been used in studies on protoplasm because of the large size of their cells and the active streaming of protoplasm that can be observed in forms such as *Nitella*; *Ulva*, *Hormidium* and particularly *Chlorella* have been repeatedly employed in experiments on assimilation; the eggs of *Fucus* have also been objects of study from various points of view, especially in reference to growth substances. In this chapter certain recent papers have been selected for a survey because of their more general interest and bearing on the life of the algae, but their scope is by no means comprehensive and they have been chosen in order to provide the student with some idea of the nature of the knowledge that is being accumulated at present. The chapter on Ecological Factors (cf. p. 349) will also be found to contain much that can be regarded as algal physiology and should therefore be consulted in this connexion.

CHLOROPHYCEAE

A recent study by Steward and Martin (1937) of the distribution and physiology of *Valonia* at the Dry Tortugas in the West Indies has brought out some interesting features. There are two species growing on the reefs that form the Tortugas; *V. macrophysa* which branches freely and *V. ventricosa* which is unbranched (cf. fig. 54). The former only grows in protected places, frequently where there is no open communication with the sea, whilst the latter grows in places exposed to the marine currents. The distribution of these two species is therefore complementary and it is suggested that they are perhaps simply ecological forms. They do, however, differ from each other biochemically in their K/Na ratio even though the Cl^- content is about the same in both species, but this is not necessarily of taxonomic significance. It is pointed out that the vesicle should not be regarded as an enormous single cell but as a fluid enclosed within a coenocytic wall composed of living cells. In contrast to the observations of many workers it was found that the enclosed fluid or sap of a *Valonia* plant in contact with sea water does not have the fixity of composition that has been ascribed to it. The sap of both species can, for most purposes, be regarded as a mixed solution of sodium and potassium chlorides, *V. macro-*

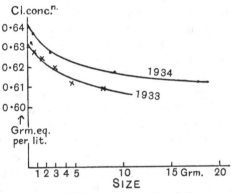

Fig. 179. Size of vesicles of *Valonia ventricosa* in relation to sap concentration. (After Steward and Martin.)

physa being poorer in potassium and richer in sodium than *V. ventricosa*. The K^+ and Na^+ content is definitely influenced by illumination; a bright light, for example, induces a high K^+ and low Na^+ content, whilst increased mechanical protection also raises the potassium and sodium chloride content, the former more so than the latter. The total salt concentration is also affected by the size of the vesicle as may be seen by a study of fig. 179.

Another interesting observation was that when the chloride concentration of sea water is increased, but not otherwise, the sap will respond to small increments of K^+ in the medium. Previously it had always been thought that changes in concentration of K^+ in the sap were related to the concentration of hydroxyl (OH^-) ions in the medium, whereas it is now evident that the relationship with the concentration of Cl⁻ in the medium is much the closer. In nature, *Valonia* derives its salts from a fluid which is much more alkaline than its sap, but unlike most other marine plants it does not appear to be able to accumulate bromide from sea water. The determining factors for these two species appear to be:

(1) Exposure to surf; a feature of the environment which operates mechanically and also through variations in pH, oxygen concentration and temperature. *V. macrophysa* is tolerant of considerable variations in the last three factors but it is not tolerant of the mechanical effects, whereas *V. ventricosa* responds in the reverse manner.

(2) Physical character of the substratum.

(3) Composition of the fluid medium, especially in respect of sodium, potassium and chloride ions.

(4) Illumination.

Rhodophyceae

Some depth studies in relation to photosynthesis by Tshudy (1934) may usefully be considered here, whilst further references to this particular problem will be found on pp. 293 and 357. Nearly all studies of photosynthesis in the algae regard Englemann's theory that the colour of the alga is complementary to that of the incident light as the basis for the investigation. For example, in the green algae the greatest assimilation takes place in the red region of the spectrum, whilst in the red algae it takes place in the green region. In 1909 Hanson, considering only those Rhodophyceae which grow at considerable depths, suggested that the chlorophyll utilized the energy that was absorbed by the phycoerythrin, in which case the red colouring matter was simply acting as a passive colour screen. This theory, however, still left unsolved the problem of the function of the coloured pigment in those red algae which always grow in the littoral belt, though it was of course possible that those algae grew in such situations merely because they could survive the competition. In 1920, Moore, Whitley and Webster showed that the Rhodophyceae assimilated less rapidly than the Chlorophyceae in bright sunlight and more rapidly than the Chlorophyceae in diffused light, and so they argued that phycoerythrin does not act as a passive colour screen but takes an active part in the process of assimilation. However, their results are partially invalidated in that no allowance was apparently made for any temperature effect.

Tshudy inserted the algae to be investigated into test tubes which were then placed horizontally in wire baskets that could be lowered to any required depth. The oxygen was measured in these tubes before and after each experiment by the Winkler method. Estimations were also made on blank controls, but as these exhibited some fluctuations the conclusions from the experiments themselves ought to be accepted with some degree of caution.

Tshudy found that:

(1) At 25 m. depth respiration takes place more rapidly than photosynthesis, but for the species investigated in that particular area there is a slight balance in favour of photosynthesis at 22·5 m.

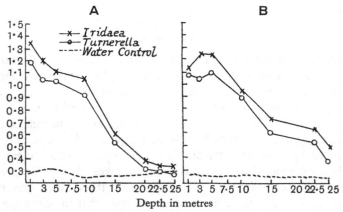

Fig. 180. Assimilation of two species of the Rhodophyceae in relation to depth and weather conditions. A, cloudy day and water slightly choppy. B, clear, calm day. (After Tshudy.)

(2) Photosynthesis is materially affected by the degree of cloudiness and the state of the water, whether it is calm or choppy. The influence of these two factors can be seen from an examination of fig. 180.

(3) On clear calm days maximum photosynthesis occurred at a depth of about 5 m., but on choppy days it occurred at or near the surface (fig. 180).

From the above results it was concluded that the red phycoerythrin acts largely as a colour screen, the plants utilizing the light in the same way as aerial shade plants. The colour of the plant would therefore seem to act in a purely physical fashion and not in any physiological manner, an interpretation that has also been supported by Seybold (1934) (cf. p. 293). It must, however, be remembered that depth studies cannot yield valid conclusions on the role of pigments unless the spectral composition of the light has been determined in order that an adequate comparison can be made with the behaviour of green algae under similar conditions.

PHAEOPHYCEAE

A study by Stocker and Holdheide (1937) of the assimilation of the principal brown fucoids which zone our shores (*Fucus platycarpus*, *F. vesiculosus*, *F. serratus*) when compared with that of a

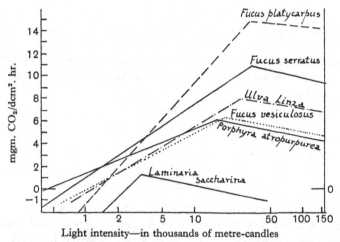

Fig. 181. Assimilation of different algae in relation to light intensity. (After Stocker and Holdheide.)

Laminaria, a green and a red alga produced some interesting results. Assimilation by all these algae decreases under very bright light, and this agrees with Tshudy's results for clear days when he found that the maximum assimilation did not take place at the surface (cf. fig. 181). The optimum light intensity for assimilation was found to be in the same region as that of the cormophytic land plants, even though these do not usually show a drop beyond 50,000 lux. The optimum temperature for photosynthesis in

Delesseria, Enteromorpha and *Fucus* is about 25° C. although there
is a fairly wide range: at low light-intensities, for example, there is
as much assimilation at 5° C. as at 15° C. (cf. fig. 182). In nature
the temperature optimum generally corresponds to the temperature
attained by the thallus in the sun's rays, whilst Ehrke (1931) also

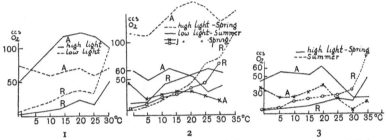

Fig. 182. Respiration and assimilation in relation to temperature and light.
A = assimilation. R = respiration. Experimental period = 3 hours. 1 = *Delesseria*.
2 = *Enteromorpha compressa*. 3 = *Fucus serratus*. (After Ehrke.)

found a correlation between the temperature of maximum assimilation and the average temperature of the month of maximum
development.

TABLE VI

Optimum temperature for assimilation	Average temperature in months of maximum development	Species
17° C.	17° C., Aug.–Sept.	*Fucus, Enteromorpha*
0° C.	0° C., winter and early spring	*Delesseria*

The principal limiting factor for assimilation appears to be the
water content because exposed thalli quickly dry up and cease to
assimilate, whilst respiration also sinks very low (cf. fig. 183). For
the Fucaceae on a normal cloudy day the amount assimilated during
the time they are exposed in 24 hours is only 0·7–1·4 % of the dry
weight, although the fertile tips of *Fucus platycarpus* acquire a
slightly higher percentage. On remoistening, *Enteromorpha Linza*
and *Porphyra umbilicalis* take up water at once and very soon
commence to assimilate again, whilst the table below shows that
the Fucaceae behave in a very different fashion.

The influence of rain on the Fuci appears to cause a reversible
depression of the assimilation rate amounting to 19–25 %, and this
factor may assume considerable importance on some coast-lines.

PHYSIOLOGY
285

Table VII. *Percentage of normal assimilation reattained*

Pelvetia canaliculata	70–80 % 8–9 hours after 1st tide; exposed for 11 days previously		
Fucus spiralis (and *F. platycarpus*)	49 % 8–9 hours after 1st tide; 3 days' exposure previously	97 %	4 hours after flooding; exposed 5 hours previously in air dryness 10 % of max., or 90 % R.H.
Fucus vesiculosus	20 % 8–9 hours after 1st tide; 3 days' exposure previously	72 %	
Fucus serratus	Cannot tolerate 3 days' exposure	42 %	
Laminaria digitata	Cannot tolerate 1–2 hours' exposure		

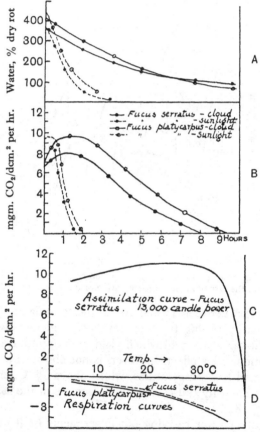

Fig. 183. A, water loss, and B, assimilation in relation to exposure (drying) on sunny and cloudy days. C, D, effect of temperature on respiration and assimilation of *Fucus*. Investigational period for assimilation, 5 min.; for respiration, 18 min. (After Stocker and Holdheide.)

The results so far described would seem to be against a weak light and cold medium as being the best conditions for the Fucaceae, but the great development of this group in northern and arctic seas cannot be overlooked, and the explanation must be the working together of factors other than merely light and temperature. Hyde (1938), however, explains this development in arctic waters as due to the indirect effect of lowering the temperature because this results in an excess of assimilation over respiration.

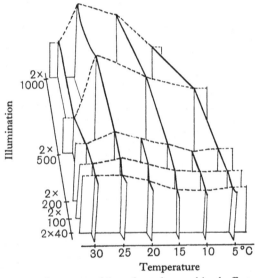

Fig. 184. Diagram of paper model to show the combined effects of light and temperature on the rate of apparent assimilation of *Fucus serratus*. (After Hyde.)

She found that between 15 and 20° C. the assimilation rate could be increased by raising the light intensity, and that there was a certain light value (2 × 500) which yielded an optimum in the rate of assimilation. This effect, however, is not observed at low light intensities and low temperatures, whilst above 25° C. an increase of the light intensity causes a marked decrease in the assimilation rate (cf. fig. 184).

Ehrke (1931), after carrying out experiments both in the field and laboratory, found that the respiration of most algae increased with rise of temperature, and often has not reached its maximum even when the high temperatures have reduced the assimilation rate considerably (cf. fig. 182). The assimilation of any alga exhibits a well-marked optimum which depends upon both light and temperature: at low temperatures the optimum occurs at low light intensities, and this is obviously significant in the case of those algae that grow in cold waters. Those algae which behave in the same way as terrestrial shade plants build up food reserves during the cold time of the year when conditions are favourable for them, and then lose the material during the more unfavourable warmer periods, whilst in summer time those algae which behave like terrestrial sun plants have a high assimilation rate (cf. p. 359).

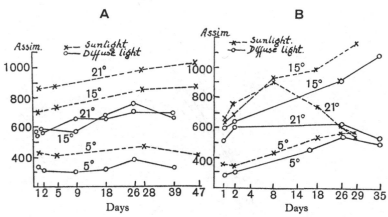

Fig. 185. Daily drift in assimilation of algae at different temperatures in sunlight and diffuse light. A, *Fucus serratus*, winter plant. B, *Porphyra*. (After Lampe.)

More recently Lampe (1935) has re-emphasized the fact that the relation between temperature and distribution is not completely solved, and furthermore he stresses the fact that it does not appear to be wholly dependent upon a physiological basis. In winter, the assimilation rate of *Fucus serratus* plants is found to rise when it is measured in sunlight under conditions of increasing temperatures (cf. also Ehrke above); on the other hand, in the case of a red alga such as *Porphyra*, when the temperature is raised above 15° C. the assimilation curve is lowered immediately in diffused light and after seven days in sunlight (cf. figs. 185, 186). From this it may be concluded that *Fucus* is an eurythermal species, i.e. tolerates a wide range of temperature, whilst *Porphyra* is a stenothermal species, i.e. tolerates only a narrow range of temperature (cf. also p. 367).

At the present time further studies are required in order to ascertain the mechanism involved in the gradual change in value of the temperature for optimum assimilation as one passes from cold to warm weather and vice versa. The explanation of this phenomenon should provide us with the clue to the correlation between algal distribution and temperature.

We may now turn to other aspects of algal physiology, and here one may mention some results of Haas and Hill (1933) who found a

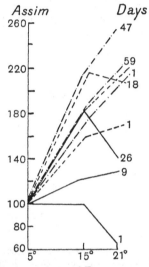

Fig. 186. Assimilation of winter plants of *Fucus serratus* at varying temperatures and in different light intensities. ———=weak diffused light, – – –=strong diffused light, – · – · =sunlight. (After Lampe.)

correlation of fat content with the vertical distribution, or in other words with the duration of exposure (cf. Table VIII).

The products of nitrogen metabolism, however, are not correlated in the same way. The algae are also characterized by an absence or paucity of free sugars, their place being taken by sugar alcohols such as mannitol. As these are probably secondary products derived from the free sugar, the latter is not to be found because conversion to a sugar alcohol removes it as fast as it is formed. It has recently been discovered, however, that the percentage of mannitol together with another substance, laminarin, in the alga *Eisenia bicyclis* tends to reach a maximum in the evening whilst the mean content for day and night is not appreciably different. In this alga, therefore, these substances are probably not direct products of photosynthesis. It has been shown that when *Eisenia* is kept in the dark both these substances decrease and so it is suggested that they should more properly be regarded as food reserve materials.

Studies on the induction of polarity in *Fucus* eggs have shown that unilateral light is normally the strongest determinant but that polarity can also be induced by an electric current or pH gradients. There is also the effect of neighbouring groups of eggs or the nearest large egg, the first rhizoid developing on the side nearest to

TABLE VIII

	Ether-extracted fat	True fat
Pelvetia canaliculata f. *libera* (salt marsh)	8·62	8·0
Pelvetia canaliculata (spray zone)	4·88	4·9
Fucus vesiculosus ecad *volubilis* (salt marsh)	3·76	
Ascophyllum nodosum (middle littoral)	2·87	
Fucus vesiculosus ⎫	2·60	2·6
Halidrys siliquosa ⎬(low littoral)	2·18	
Himanthalia lorea ⎭	1·21	
Desmarestia aculeata ⎫	0·65	
Laminaria digitata ⎪	0·46	0·3
Pelagophycus ⎪	0·27	
Macrocystis ⎬(sublittoral)	0·34–0·40	
Nereocystis ⎪	1·06	
Laminaria Andersonii ⎭	0·65	

them. This is probably a growth substance effect since the presence of growth substances has been demonstrated in eggs, sperm and fruiting tips of *Fucus*. In the case of polarity induced by the proximity of eggs the growth substances from the neighbouring ova are not *necessary* for rhizoid formation but are purely directive.

GENERAL

An investigation has been carried out by Biebl (1938) using hypo- and hypertonic solutions of sea water for determining the drought resistance and the osmotic relations of algae from different depths. The algae studied could be placed in three groups according to their behaviour but irrespective of their geographical locality.

A. Deep growing algae which are never exposed to the air: they are resistant up to a concentration of 1·4 times sea water.

B. Algae of low water mark and the lower littoral tide-pools which rarely become completely dry: these are resistant up to concentrations of 2·2 times sea water.

C. Algae of the littoral belt: these are usually completely exposed and can resist a concentration of 3·0 times that of sea water.

The behaviour of these algae is summarized in Table IX.

It was found that most of the Rhodophyceae possess a cell sap which has an osmotic pressure approaching that of the maximum hypertonic resistance likely to be encountered in their habitat, but this correlation between cell sap and external medium is not so evident in the case of the Chlorophyceae and Phaeophyceae. In their resistance to desiccation the algae fall into the same three ecological groups as can be seen from Table X.

In group 1 when the filament is dried up for only a very short time by means of filter paper the cells die or collapse so quickly that they do not even recover when put back into sea water. Those of group 2 are less susceptible and those of group 3 hardly susceptible to this treatment.

A study of the chlorophyll relations in all three algal groups by Seybold and Egle (1938) revealed the fact that only in the Chlorophyceae are both chlorophylls a and b present, their proportions being the same as those in the submerged flowering plants. In the Rhodophyceae, Phaeophyceae, Cyanophyceae and Bacillariophyceae only chlorophyll a is present, so that if the absence of chlorophyll b is considered to represent a primitive character, the Chlorophyceae would have to be regarded as the most recent group (cf. p. 265). On the other hand it is equally possible that in these groups the second component has been lost, possibly due to the introduction of the extra colouring pigment or to some other factor. In not one of these groups does there appear to be any relation between depth and quantity of chlorophyll and carotene present, the actual amount being determined rather by the genetic constitution. The quantity of pigment per dry or fresh weight is less in the Rhodophyceae and Phaeophyceae than it is in the Chlorophyceae, but the fact that members of the first two groups assimilate carbon dioxide as rapidly, weight for weight, as those of the Chlorophyceae indicates that their carbon assimilatory apparatus cannot be deficient. It is also evident that the green algae exhibit a far greater range in the amount of pigment present than do the red and brown algae (cf. fig. 187).

TABLE IX. *Osmotic resistance, drought resistance and habitat of some Heligoland marine algae*

	Osmotic resistance. Range during 24 hours = Range of sea water concentration	Drought resistance after 13 hours in a damp chamber over NaCl solutions		Habitat depth
		Plants still living over g. NaCl in solution =	Relative humidity %	
A. *Antithamnion plumula*	0·5–1·5	0·5	98·4	8–16 m. deep
Trailliaella intricata	0·5–1·4	0·5	98·4	,,
Plocamium coccineum	0·5–1·3	0·4	98·8	,,
Halarachnion ligulatum	0·5–1·5	1·0	96·8	,,
Brogniartiella byssoides	0·4–1·6	Dead already over sea water	—	,,
B. *Dictyota dichotoma*	0·7–1·7	1·6	94·6	At low-water mark.
Membranoptera alata	0·4–1·7	1·6	94·6	Mostly under water.
Ptilota elegans	0·5–2·2	1·6	94·6	,,
Polysiphonia urceolata	0·2–2·1	Dead already over sea water	—	,,
Polysiphonia nigrescens var.	0·2–2·2		—	,,
Polysiphonia nigrescens var.	0·1–2·2		—	,,
C. *Rhodochorton floridulum*	0·3–2·1	3·0	88·0	Littoral zone
Porphyra laciniata	0·2– ⎱ Over			,,
Ulva lactuca	0·2– ⎰ 3·0	Above 4·0	Under 83	,,
Ulva Linza	0·4–			,,
Cladophora rupestris	0·1–			,,
Cladophora gracilis	0·2–2·1	3·4	86·0	,,
Elachista fucicola	0·2–2·4	3·4	86·0	,,

TABLE X. *Resistance to drought in air of different dampness. The three ecological groups are sharply distinguished by their behaviour*

NaCl g.:	0·3	0·4	0·5	0·6	0·7	0·8	0·9	1·0	1·1	1·2	1·6	1·7	1·8	2·0	2·2	2·4	2·6	3·0	3·4	3·6	4·0	Habitat
% R.H.	99·1	98·8	98·4	98·1	97·8	97·5	97·2	96·8	96·4	96·1	94·6	94·2	93·8	93·0	92·2	91·1	90·6	88	86	85	83	
O.P. (atm.) at +20° C.	12·4	16·6	20·8	25·2	29·5	34·0	38·5	43·0	47·6	52·2	73·0	78·5	84·0	96·0	108·5	122·0	136·0	166·0	208·3	215·2	249·0	
Plocamium coccineum	i		o	o	×	×	×	×	×	×												I. Depth algae
Antithamnion plumula	i	i	i	o	o	o	×	×	×	×												
Trailliella intricata	i	i	i	o	o	×	×	×	×	×												
Halarachnion ligulatum	i	i	i	i	i	i	i	i	o	o	o			×								
Polysiphonia nigrescens	×	×	×	×	×	×	×	×	×	×				×								II. Algae of low-tide mark
Membranoptera alata										i	i		o	o	o	o						
Ptilota elegans										i	i		o	o		o	×					
Dictyota dichotoma										i	i		×	×	×	×	×	×				
Porphyra laciniata														i	i	i	i	i	i	i	i	III. Littoral algae
Rhodochorton floridulum														i	i	i	i	i	o	o	×	
Elachista fucicola														i	i	i	i	i	i	o	×	
Ulva lactuca														i	i	i	i	i	i	i	i	
Ulva Linza														i	i	i	i	i	i	i	i	
Cladophora rupestris														i	i	i	i	i	i	o		
Cladophora gracilis																				o	c	

i, living; o, partly dead; ×, wholly dead.

The light relations in the photosynthetic mechanism of the algae can be divided into two components:

(*a*) Physical component, which is the amount of light energy absorbed by the thallus.

(*b*) Physiological component, which is the amount of absorbed light energy that is actually employed in carbon assimilation.

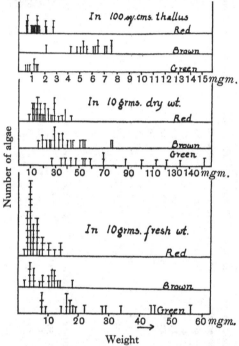

Fig. 187. Chlorophyll content of the same number of red, green and brown algae expressed in different terms. (After Seybold and Egle.)

The light energy relations of selected members in the Chlorophyceae, Phaeophyceae and Rhodophyceae are illustrated in Table XI from Seybold (1934) and also in fig. 188.

It is at once obvious that at depths below 1 m. the Rhodophyceae and Phaeophyceae are much more efficient as metabolic machines than the Chlorophyceae, and several species are even more efficient at the surface. As a result of his studies Seybold (1934) concluded that Englemann's theory of complementary colours is only valid for the physical component of the light relationship, that is, the

TABLE XI

Depth in in.	Chlorophyceae			Rhodophyceae				Phaeophyceae	Light intensity in % of incident light
	Monostroma	Ulva	Enteromorpha	Delesseria	Porphyra	Phyllophora	Chondrus	Laminaria	

Percentage absorption of white light at different depths

Depth in in.	Monostroma	Ulva	Enteromorpha	Delesseria	Porphyra	Phyllophora	Chondrus	Laminaria	Light intensity in % of incident light
0	39	46	59	33	57	58	80	78	100
1	29	36	40	27	42	43	56	56	82·3
5	14·9	19	22	16	26	24	30	27	44·6
10	7·4	9	12	9·2	13	12·8	16	16	24·0
20	2·2	2·8	3·4	2·8	4	4·3	4·7	4·6	7·5
30	0·9	1	1·4	1·1	2·3	2·2	2·7	2·7	3·0
40	0·54	0·54	0·8	0·73	1·3	1·4	1·6	1·6	2·0
50	0·3	0·3	0·42	0·55	0·75	0·8	0·9	0·9	1·5

Percentage absorption of light available at the different depths

Depth in in.	Monostroma	Ulva	Enteromorpha	Delesseria	Porphyra	Phyllophora	Chondrus	Laminaria
0	39	46	59	33	57	58	80	78
1	35	44	49	33	40	52	69	69
5	33	42	49	36	58	54	68	61
10	31	38	50	38	54	53	67	67
20	30	37	45	37	53	57	63	63
30	30	33	47	37	75	73	90	90
40	30	30	45	40	75	77	95	95
50	30	30	42	55	75	80	95	95

pigments only help in the amount of light absorbed and not in its utilization. We have already seen earlier that other workers, who have studied members of the individual groups, have arrived at a similar conclusion since they regard the coloured pigments as acting in the same way as a colour screen. Apart from the physical adaptation, in the sense of complementary light absorption, there is

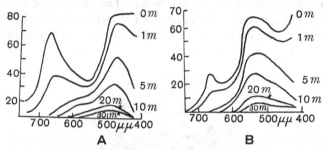

Fig. 188. Absorption curves of A, *Monostroma*, B, *Delesseria*, at different depths. (After Seybold.)

also a physiological adaptation to strong and weak light and to long and short wave light. The algae can be placed into two groups depending upon their responses to strong and weak light, or to long wave and short wave light. This problem, however, is followed up more closely in the later chapter on algal ecology (cf. p. 359).

*SYMBIOSIS

The most striking and well-known examples of symbiosis in the algae are provided in those cases where the plants are associated with animals, especially Coelenterates, or with fungal threads, as in the common lichens. Apart from these examples, however, there are other cases which are not so well known, largely because they are not so common or so conspicuous. *Gloeochaete*, for example, is a colourless genus of the Tetrasporaceae which possesses blue green bodies that look like chromatophores, though they are really a symbiotic blue-green alga. *Gleucocystis* is a colourless genus of the Chlorococcales in which a symbiotic member of the Cyanophyceae also forms blue green "chromatophores" that appear as a number of curved bands grouped in a radiating manner around the nucleus. In this case the illusion is further enhanced because they break up into short rods at cell division. It has so far proved impossible to grow the blue-green alga separately and it may thus have lost its power of independent growth. *Geosiphon*, which is variously regarded as a Siphonaceous alga or as a Phycomycete, possesses small colonies of *Nostoc* enclosed in the colourless pear-shaped vesicles that arise from an underground weft of rhizoidal threads. Reproduction by the formation of new vesicles is said to occur only in the presence of the *Nostoc*. The presence of chitinous material in the vesicular wall suggests a fungal nature for *Geosiphon*, the vesicles perhaps being galls that are formed on the threads as a result of the presence of the alga.

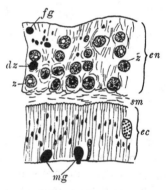

Fig. 189. Symbiosis. Zooxanthellae in the tissues of a coelenterate, *Pocillopora bulbosa* (× 375). *ec*=ectoderm, *en*=endodermis, *dz*=dead zooxanthellae, *fg*=fat globule, *mg*=glands, *sm*=structureless lamella, *z*=zooxanthellae. (After Yonge.)

The principal genera taking part in lichen synthesis are *Nostoc*, *Scytonema*, *Cystococcus*, *Gloeocapsa* and *Trentepohlia*. To what extent the lichen body is a case of true symbiosis is a problem that is still not wholly settled: under normal conditions it is probably a real symbiotic relationship but under abnormal conditions the fungus may become a parasite and devour the algal component. The green bodies which are found associated with the cells of Coelenterates and Radiolarians are usually placed in what may be called a "form" genus, *Zooxanthella* (cf. fig. 189). Most of the species belong to the Cryptophyceae, but in certain of the Coelenterata the motile phases of some of the algae which have been discovered suggest an affinity to the Dinophyceae, whilst *Chlorella* (Chloro-coccales) is also said to behave as a symbiont of this type. The non-motile vegetative cells are usually found in the peripheral layers of the polyp, the larval stages of the host commonly being devoid of the alga. Most of the algal symbionts are known to have a motile stage and hence are capable of an independent existence. The function and relations of these symbiotic algae in the coral polyps has been discussed at great length by Yonge (1932), and on the whole there would appear to be evidence for a symbiotic relationship, the alga obtaining food from the animal, and the animal oxygen and also perhaps nitrogen from the alga. The problem of the relationships between algae and animals is by no means completely worked out, and it is not impossible that in some cases we really have an animal that is parasitizing the alga. This is probably especially true in the case of the worm-like creature *Convoluta Roscoffensis* and its algal associate *Carteria*, because the animal apparently cannot live unless infected with the alga, whilst under certain conditions it also devours the green cells.

Other examples of symbiosis are provided by *Anabaena Cycadearum* which lives in the root tubercles of species of *Cycas*, and *Anabaena Azollae*, which is found in the leaves of the water fern, *Azolla filiculoides*, though the species of *Nostoc* that are to be found in the thallus of the Liverworts *Blasia* and *Anthoceros* are probably no more than space parasites obtaining shelter.

Epiphytism is extremely common among the algae, whilst there are also a number of epizoic forms. One may also find endophytic species, such as *Schmitziella mirabilis* in *Cladophora pellucida*, and endozoic species, such as *Rhodochorton endozoicum* in the sheaths of hydroids. The origin of the symbiotic habit among the algae is probably to be explained as cases of epiphytism in which the relationship between host and epiphyte became more intimate: similarly the relatively few cases of parasitism probably arose either directly from an epiphytic habit or else passed through the symbiotic phase. Examples of total or partial parasites are *Notheia anomala* in the Phaeophyceae, *Choreocolax Polysiphoneae* in the Rhodophyceae and *Chlorochytrium Lemnae* in the Chlorophyceae.

SOIL ALGAE

Terrestrial algae may be classified conveniently as follows:

(1) Aero-terrestrial species found growing on plants.
(2) Eu-terrestrial.

>   True soil species:
>
>   (a) Epiterranean, or lying in the surface layers of the soil.
>   (b) Subterranean, or lying in the lower layers of the soil. So far as is known at present there are no obligate species of this class.
>   (c) Hydroterrestrial, or occupying the soil of aquatic areas.
>   (d) Casuals.

The study of soil algae, as such, began seriously at the commencement of the nineteenth century with the works of Vaucher, Dillwyn, Agardh and Lyngbye, whilst towards the end of the century monographs by Bornet and Flahault, Gomont, Wille and the Wests, father and son, began to make their appearance. In 1895 Graebner, in a study of the heaths of North Germany, gave the first account of soil algae as ecological constituents, and subsequently many ecologists have shown that soil algae are pioneers on bare soil where they prepare the ground for the higher plants that follow. In such cases the algal flora is generally richest when the soil is primarily or secondarily naked, e.g. mud flats developing to salt marsh, or ploughed grassland. A manured soil also has a very rich flora, whilst the same species are to be found in unmanured soils, though not in such numbers. The richness of the flora is also influenced by the moisture conditions, damp soils having a more varied and extensive collection of algae than dry soils. In recent years dilution cultures have been widely used in order to give a quantitative aspect to the work, and the results of such studies have been to show that there is probably a seasonal variation in numbers, but that the behaviour depends on the depth and kind of soil.

Subterranean Algae

There are great fluctuations in the numbers of the different species that compose the flora, but there are no species in the lower layers of the soil which do not also occur in the surface layers. Dilution cultures, together with the counting of samples, have

shown that the algal flora is mainly confined to the top 12 in. of
soil with a maximum abundance at about 3–6 in. below the surface.
With increasing depth the number of algae decrease regularly, the
maximum depth at which they have been recorded being two
metres; there is, however, really no conclusive evidence which
shows that algae can grow in the deeper layers where there is no
light, and it is very probable that they are only present in these
layers in a resting phase. The number of reproductive bodies in the
surface layers reaches a maximum in spring, but in the lower levels
it remains constant throughout the year. In Denmark the quality
of the soil is apparently decisive in determining the luxuriance of
the flora irrespective of whether the ground has been disturbed or
not. In Greenland soil algae have been found down to a depth of
40 cm., and their presence there can only be satisfactorily explained
by the action of water trickling down the cracks because burrowing
animals are absent. A study of soils from all over the world has
emphasized the existence of a widely distributed algal formation in
cultivated soils. This flora consists of about twenty species of
diatoms, twenty-four of Cyanophyceae and twenty species of green
algae, among which *Hantzschia amphroxys*, *Trochiscia aspera*,
Chlorococcum humicolum, *Bumilleria exilis* and *Ulothrix subtilis* var.
variabilis are the most frequent.

The growth of these soil algae has been a source of interest and
experiment for a number of years. Roach (1926) has found that
ordinary growth in *Scenedesmus costulatus* var. *chlorelloides* is best
in a glucose medium but that xylose is toxic, the factors controlling
the normal growth rate being light, temperature and aeration of
the medium (cf. fig. 190). The same alga has been used for growth
experiments in the dark in order to determine how far such algae
can grow when they are below the soil surface (cf. fig. 191). This
and four other species can be made to grow in the dark provided an
organic medium is present, but they all react differently to the
various conditions and also they vary in the amount of growth that
occurs. At constant temperature, increasing the light intensity
from $\frac{1}{25}$ to $\frac{1}{4}$ has a far greater effect than increasing the intensity
from $\frac{1}{4}$ to full sunlight. Under full light the growth curves (cf. fig.
191) rise to an optimum by means of photosynthesis alone, but at
lower intensities the optimum is only approached if additional
nutriment, in the form of glucose, is present as well. There is no

adequate evidence that such organic media are present in the soil layers so that it is very doubtful whether growth in the dark can occur in nature, but it has been shown, however, that *Nostoc punctiforme* from the leaves of *Gunnera* and also a species of *Euglena* are capable of growth in the dark.

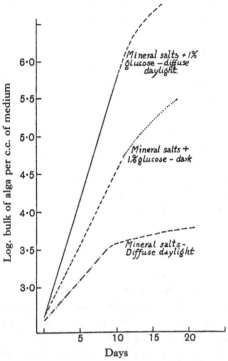

Fig. 190. Growth of the soil alga, *Scenedesmus*, under different conditions of nutrition and light. (After Roach.)

Even if the algae cannot grow in the lower layers of the soil because of the darkness, we must still enquire into the process responsible for their appearance in those layers. The possible agencies are (1) cultivation, (2) animals, (3) water seepage and (4) self-motility. Mechanical resistance and lack of light are said to prevent the Cyanophyceae from moving down under their own locomotion, and whilst it is possible that algae may move down through their own motility, further experimental work on this aspect is much to be desired. The effect of water seepage will depend on the heaviness of the rainfall, the state of the soil, i.e. whether dry and cracked, and the nature of the algae, i.e. whether or not they possess a mucous sheath. Passage through the soil is facilitated in the filamentous algae either by fragmentation or else by the formation of zoospores, the factors that are responsible for the former process appearing to differ for the various species. Many

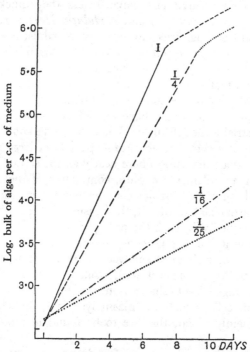

Fig. 191. Rate of growth of *Scenedesmus* in a solution of mineral salts under different light intensities. (After Roach.)

green algae are known to form zoospores when put into water after a period of dryness, and hence one may presume that a shower of rain will also induce zoospore formation. Petersen (1935) has demonstrated experimentally that rain can carry algae down efficiently to a depth of 20 cm., but that the process is facilitated by the presence of earth-worms, although these animals probably only operate indirectly in that they loosen the earth. Farmers in the course of their cultivating operations must frequently be responsible for the conveyance of algae down into the soil.

Many of the soil algae, especially the Cyanophyceae, can resist very protracted spells of dryness as Roach (1920) demonstrated when soils from Rothamsted that had been kept for many years were remoistened. Bacteria developed first, then unicellular green algae with some occasional moss protonemata, and although the Cyanophyceae appeared last, nevertheless they quickly became dominant. *Nostoc muscorum* and *Nodularia Harveyana* appeared after the soil had been dried up for 79 years, whilst *Nostoc Passerinianum* and *Anabaena oscillarioides* var. *terrestris* appeared after 59 years of dryness. These algae differed in some respects from the typical forms that are to be found in ordinary soils, but this was probably only due to the cultural conditions.

Fritsch and Haines (1922) have studied the moisture relations of some terrestrial algae (cf. fig. 192) and they have shown that:

(1) There is a complete absence or paucity of large vacuoles.

(2) In an open dry atmosphere nearly all the sap is retained.

(3) When the filaments dry up, contraction of the cell is such that the cell wall either remains completely investing the protoplast or else in partial contact with it, thus ensuring that all the moisture which is imbibed will reach the protoplast.

(4) During a drought there is, as time goes on, a decreasing tendency for the cells to plasmolyse and there are also changes in the permeability of the cell wall, whilst the access of moisture normally brings about changes in the reverse direction. The majority of cells which do plasmolyse lack the characteristic granules, mainly of fat, that are to be found in most terrestrial algae.

(5) Those cells which survive after drought do not contain any vacuoles and possess instead a rigid, highly viscous protoplast which is in a gel condition. This is the normal state of the vegetative cells of *Pleurococcus* and the cells in the "*Hormidium*" stage of *Prasiola*.

(6) If desiccation is rapid most of the cells will die but some will plasmolyse and retain their vitality in that state for weeks or months. In spite of the death of the bulk of the cells no species disappears from the flora during a rapid onset of drought.

(7) If desiccation continues, the number of living resting cells will remain constant for several years.

(8) During a very long drought the resting cells of algae below the surface will still survive.

Apart from the moisture relations there are also other factors that may be involved. Diatoms can survive very low temperatures,

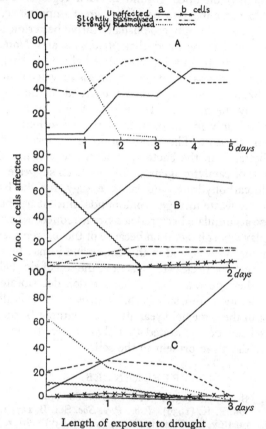

Fig. 192. Effect of exposure to drought on A, *Hormidium*; B, "*Hormidium*" stage of *Prasiola*; C, *Zygogonium ericetorum*. (After Fritsch and Haines.)

−80° C. for 8 days or −192° C. for 13 hours, whilst dry spores of *Nostoc* sp. and *Oscillatoria brevis* can survive −80° C., though if they are moist a temperature lower than −16° C. will kill them. As the vegetative filaments of *Nostoc* die after four days at −2 to −8° C. this genus must survive severe winters in the form of spores. So far as the algae of tropical soils are concerned the dry spores of *Nostoc* sp. and *O. brevis* can tolerate 2 min. at 100° C., the wet spores 20 min. at 60–70° C., and the vegetative filaments 10 min. at 40° C., this latter being a temperature that is frequently reached on open ground in such regions. Acidity and alkalinity do not appear to be of any great importance, although members of the Chlorophyceae usually thrive better on basic soils.

It has been demonstrated that *Anabaena* and *Nostoc* can fix nitrogen from the air in the presence of light, but other soil algae apparently do not possess this power unless they occur in combination with bacteria, and even then the actual fixation is probably carried out by the bacteria. It has been found by De (1939) that *Anabaena* will only fix nitrogen from the air so long as nitrate is absent from the soil. The combination of bacteria and algae fix nitrogen better than the bacteria do alone, so that the algae must act as a kind of catalytic agent, and it has been suggested that they (a) provide carbohydrate, and hence energy, for the bacteria, or (b) remove the waste nitrogen compounds, since it has been shown that if these accumulate bacterial activity is reduced. In some cases the algae play a part in aeration because of the oxygen they produce during photosynthesis, and in this connexion it may be mentioned that unless certain species are present in the soil of rice fields during the period they are waterlogged the aeration deteriorates and the rice becomes much more susceptible to disease. Rice is also capable of growing in the same field year after year without being manured, and it has been demonstrated that this is due to the fixation of nitrogen by the algae present in the soil.

REFERENCES

Physiology.   BIEBL, R. (1938). *Jb. Wiss. Bot.* **86**, 350.
Soil Algae.   DE, P. K. (1939). *Proc. Roy. Soc.* Ser. B, **127**, 121.
Physiology.   DU BUY, H. G. and OLSON, R. A. (1937). *Amer. J. Bot.* **24**, 609.
Physiology.   EHRKE, G. (1931). *Planta*, **13**, 221.
Soil Algae.   FRITSCH, F. E. (1936). *Essays in Geobotany in honor of W. A. Setchell*, p. 195. Univ. California Press.
Soil Algae.   FRITSCH, F. E. and HAINES, F. M. (1922, 1923). *Ann. Bot.* **36**, 1; **37**, 683.
Physiology.   HAAS, P. and HILL, T. G. (1933). *Ann. Bot., Lond.*, **47**, 55.
Physiology.   HANSON, E. K. (1909). *New Phytol.* **8**, 337.
Physiology.   HYDE, M. B. (1938). *J. Ecol.* **26**, 118.

SOIL ALGAE

Symbiosis. KEEBLE, F. and GAMBLE, F. W. (1907). *Quart. J. Micr. Sci.* **51**, 167.

Physiology. LAMPE, H. (1935). *Protoplasma*, **23**, 543.

Physiology. MOORE, B., WHITLEY, E. and WEBSTER, T. A. (1920). *Ann. Rep. Oceanog. Univ. Liverpool*, **36**, 32.

Physiology. NISIZAWA, N. (1938). *Sci. Rep. Tokyo Bunrika Daig.* **3**, 289.

Soil Algae. PETERSEN, J. B. (1935). *Dansk bot. Ark.* **8**, 1.

Soil Algae. ROACH, B. M. (1919). *New Phytol.* **18**, 92.

Soil Algae. ROACH, B. M. (1920). *Ann. Bot., Lond.*, **34**, 35.

Soil Algae. ROACH, B. M. (1926). *Ann. Bot., Lond.*, **40**, 149.

Physiology. SEYBOLD, A. (1934). *Jb. wiss. Bot.* **79**, 593.

Physiology. SEYBOLD, A. and EGLE, K. (1938). *Jb. wiss. Bot.* **86**, 50.

Valonia. STEWARD, F. C. and MARTIN, J. C. (1937). *Publ. Carneg. Instn*, no. 475, p. 89.

Physiology. STOCKER, O. and HOLDHEIDE, W. (1938). *Z. Bot.* **32**, 1.

Physiology. TSHUDY, H. (1934). *Amer. J. Bot.* **21**, 546.

Symbiosis. YONGE, C. M. and NICHOLLS, A. G. (1932). *Reports of the Gt Barrier Reef Exp.* **1**, 135. Brit. Mus. Publ.

CHAPTER XI

MARINE ECOLOGY

The algae of the rocky coasts have attracted more investigators than those of the salt-marsh coast, probably because of the greater abundance of species, the greater ease in identifying the component members of the flora, and the well-marked zonation which is so characteristic of most rocky shores. In spite of these numerous investigations we are still very far from understanding how the zonation is secured and maintained, nor is there sufficient data about the environmental factors because most workers have simply contented themselves with describing zonations in particular areas and only suggesting possible controlling factors. Furthermore, our knowledge of recolonization on the sea-shore is very rudimentary,[*] and it is highly desirable that more information should be obtained

TABLE XII. *Algal associations*

Dover	Isle of Wight	Peveril Point (Dorset)	Wembury (Dorset)	Lough Ine (Ireland)
Endoderma	—	—	—	*Lichina*
Rivularia-Calothrix	—	*Rivularia-Calothrix*	—	*Hildenbrandtia-Verrucaria-Ralfsia*
Schizothrix Fritschii	—	—	—	Upper Chlorophyceae
Enteromorpha-Urospora-Codiolum	—	—	—	—
Chrysophyceae-Endoderma-Lyngbya	—	—	*Pelvetia*	—
Chrysotila stipitata	*Fucus ceranoides*	*Fucus spiralis*	*Fucus spiralis*	*Fucus spiralis*
	Ascophyllum	*Fucus vesiculosus*	—	*Fucus vesiculosus* var. *evesiculosus*
Fucus spp.	*Fucus vesiculosus*	—	*Ascophyllum*	*Ascophyllum*
Gelidium-Polysiphonia	*Fucus serratus*	*Fucus serratus*	*Fucus serratus*	*Fucus serratus*
Ralfsia	—	—	—	—
Enteromorpha-Porphyra	—	*Porphyra*	—	*Porphyra*
Chalk-boring algae	—	—	—	*Bangia-Urospora*
Rhizoclonium-Vaucheria	—	*Ulva*	*Lomentaria*⎫	*Lomentaria*
Pylaiella littoralis	—	*Laurencia-Corallina*	*Gigartina*⎬	*Laurencia*
Enteromorpha intestinalis	—	—	—	*Gigartina*
Rhizoclonium riparium	—	—	—	*Callithamnion-Ceramium*
	Halidrys	—	—	*Nemalion*
				Himanthalia
				Corallina-Lithothamnion
	Laminaria	*Laminaria*	*Laminaria*	*Laminaria*
SUBLITTORAL				Sublittoral Rhodophyceae
				Alaria
				Plumaria-Ceramium
				Laurencia-Gelidium
SHELTERED COASTS			*Chondrus*	*Chondrus*
				Cladophora rupestris
				Lower Chlorophyceae.
				(*Enteromorpha* spp.)

[*] Cf. Rees (1940) upon recolonization.

about the factors that cause removal of algae from rocks. Statistical analyses of drift show that the majority of Laminariaceae are torn in their entirety from off the rocks, so that removal in their case cannot be due to epiphytes or to the boring of the stipe by the mollusc, *Patina pellucida*, and as they usually grow beneath low-tide mark surf action is also removed as a possible destructive factor. It may be that the continual swell and strong currents finally bring about their destruction. In the case of smaller algae, however, the weight and resistance of an excessive epiphytic flora brings about the uprooting of the host plant. This, and numerous other problems, await the attention of future investigators.

Table XII contains in a summarized form the principal communities that have been recognized around the coasts of Great Britain. It is not proposed that any of these should be described in detail, but it is hoped that the outline provided by this table may be a guide to students who visit any of these areas. One of the principal characteristics of any rocky shore is the way in which the different algae are distributed in zones or belts at the different

of the British Isles

Clare Island (Ireland)	Castletown (I.O.M.)	Cromer (Chapman, *J. Linn. Soc.* (*Bot.*), 1917)	Cumbrae (Scotland)
Lichina	—	—	—
Hildenbrandtia-Verrucaria			
Prasiola stipita			—
Enteromorpha intestinalis	—	Enteromorpha	Enteromorpha intestinalis
Pelvetia	Pelvetia	Fucus platycarpus	Porphyra-Urospora-Ulothrix
Fucus spiralis	Fucus spiralis	—	Pelvetia
Fucus vesiculosus var. evesiculosus	Ascophyllum		Fucus spiralis
—	Fucus vesiculosus	Fucus vesiculosus var. evesiculosus	Ascophyllum
Fucus serratus	Fucus serratus	Fucus serratus	Fucus vesiculosus
			Fucus serratus
Porphyra	—	Fucus-Porphyra-Enteromorpha	
Bangia-Urospora-Ulothrix	Porphyra-Urospora-Ulothrix		
Rhodymenia	Laurencia-Cladophora-Rhodochorton		
Laurencia-Gigartina	Laurencia-Lomentaria	Laurencia pinnatifida	Laurencia
	—	—	Gigartina-Cladophora
Callithamnion arbuscula			Enteromorpha Linza
Nemalion			
Himanthalia	Himanthalia	Hildenbrandtia-Lithothamnion	
Corallina			
Laminaria	Laminaria	—	Laminaria
Encrusting algae		SUBLITTORAL	
Cystoseira			
Ascophyllum		SHELTERED COASTS	
Corallina-Lithothamnion			
Corallina-Cladostephus			
Rhodochorton floridulum	Enteromorpha-Cladophora-Chordaria		

heights. On the whole, any one species usually occupies a very definite vertical range and only occasionally is to be found outside it, and then there is often some cause, such as the presence of a rock pool, in which conditions for its existence are favourable. It is not intended in this chapter to enter into any detailed discussion as to the causes or factors controlling this zonation, an aspect which is dealt with more fully in the last chapter (cf. p. 351). It is sufficient here to point out that these zonations do exist and are characteristic of a rocky shore. Furthermore, a glance at Table XII will show that on the whole the zonation is remarkably similar around most of the British Isles, and the same or very similar communities can be found at much the same level at the different localities. The actual number of communities recognized depends upon two factors:

(*a*) *The locality.* It will be observed that the two Irish stations have a much richer zonation, and this can probably be associated with their position in relation to the Gulf Stream because this will tend to produce a mixture of species from both cold and warm waters.

(*b*) *The personal factor.* Each investigator will tend to have a somewhat different concept of what is represented by an algal community, whilst the number of communities recognized will also depend upon the time and thoroughness with which the shore is examined.

The terminology that has been employed has led to no little confusion. Algal ecology, as such, commenced later than the ecology of land vegetation. Some investigators have attempted to apply the terms used in land ecology to algal ecology, whilst others have considered that the conditions are sufficiently different to make this application impossible. Cotton (1912), for example, recognized five algal formations at Clare Island:

(1) Rocky shore formation.
(2) Sand and sandy mud formation.
(3) Salt marsh formation.
(4) River mouth formation.
(5) Brackish bay formation.

These were subdivided into associations, the rocky shore formation containing the associations of the exposed coast and the associations of the sheltered coast. Although the term "association" was applied to these communities, it is probable that many of them are really mere "societies" in strict ecological nomenclature because they are only transient. At Lough Ine Rees (1935) classified the formations on a different basis and he recognized only two, the open and sheltered coast formations. Cotton's formations were based on substrate or salinity whilst Rees's were based on shelter. Rees further used the term "association" for those communities where species that are associated with the dominants are controlled by the same factors. The difficulty of this criterion is the time involved in proving experimentally that certain factors do control the distribution of the species concerned. Seasonal communities, or those which were locally dominant, were regarded as societies, whilst the term "zone" was used for those algal belts which possess horizontal continuity with well-marked upper and lower limits.

In a study of some New Zealand littoral vegetation Cranwell and Moore (1938) termed the associations of the successive belts which follow one another in a regularly recurring sequence as an "association-complex". The horizontal belts were commonly continuous but they could be interrupted occasionally by another community, e.g. one could have an association fragment of *Durvillea* in the *Xiphophora* belt. It is apparent therefore that there is some divergence of opinion about nomenclature, and at present, until a thorough resurvey of the whole problem has been carried out, it would perhaps be more satisfactory to use a non-committal term such as "community" which implies no particular status.

THE BASIC ZONATION

Out of the wealth of material available it is apparent that there is on British coasts what one may term a basic zonation, principally composed of fucoids, and on this other communities are superimposed, the actual number being dependent upon the two factors already mentioned. This basic zonation is briefly as follows:

(1) An upper *Enteromorpha* belt. Such a belt has been recorded from all the localities except those around Dorset and at Castletown in the Isle of Man. At Dover there are other species associated with the *Enteromorpha*, e.g. *Urospora* and *Codiolum*. On any coast there will be a development of an *Enteromorpha* community

wherever trickles of fresh water run down over the rocks to the sea, and it is to be supposed that the lowered salinity is responsible for this development.

(2) A zone of *Pelvetia canaliculata* can be found on most shores at about high-water mark and extending up as far as the spray goes.

(3) Immediately below this there is often a zone of *Fucus spiralis* or *F. platycarpus*.

(4, 5) The next two belts vary in position, *Ascophyllum nodosum* sometimes being the uppermost and in other places *Fucus vesiculosus*. Where both belts are present there is an intermediate zone in which the two are mixed.

(6) The lowest fucoid zone is commonly dominated by *F. serratus*, but in certain areas it may merge at low-water mark into an

(7) *Himanthalia* zone.

At the same level as the *Fucus serratus* belt one may find that it is partially replaced by communities of red algae, or that there is a zone of such communities between the *Fucus* and *Himanthalia* belts. There are three communities of this type which may be frequently encountered in the different localities:

(8) A *Porphyra* community with which *Bangia* and *Urospora* are often associated.

(9) A *Laurencia* community, the existence of which is frequently marked in summer by the development of epiphytic forms such as *Cladophora* and *Lomentaria*.

(10) A *Gigartina* community.

On sheltered coasts *Chrondrus crispus* may occur at these low levels. In the sublittoral there is commonly a bed of *Laminaria* species in which *L. digitata* tends to be dominant near low-water mark and *L. Cloustoni* farther down. *L. saccharina* appears in those areas where the substrate is more or less sandy.

The effect of the height of tidal rise upon the vertical extent of the zonations is illustrated very well in Table XIII in which the algal zones from four localities are all reduced to levels based on mean low-water mark. The small range at Bembridge and Peveril Point has resulted in a compression and overlapping of the zones, whereas at Castletown, where the range is large, the zones overlap but little and occupy a considerable vertical height. The level of the upper zones in any locality is not entirely dependent upon the

height of the spring tides. On an exposed coast the shore is open to
considerable wave action and a heavy spray dashes against the rocks
to a height of several feet. As a result of this wave action the upper

TABLE XIII

Feet above M.L.W.	Bembridge I.O.W.	Peveril Point Dorset	Castletown I.O.M.	Cumbrae Firth of Clyde
13	—	—	Porphyra	—
12	—	—	Porphyra	—
11	—	—	Porphyra	—
10	—	—	Pelvetia Ascophyllum	Enteromorpha (fresh-water drainage)
9	—	—	Pelvetia Fucus spiralis	Porphyra
8	Fucus ceranoides	Cyanophyceae	Ascophyllum Fucus spiralis Ascophyllum	Pelvetia
7	Fucus ceranoides Ascophyllum	Cyanophyceae	Fucus vesiculosus	Pelvetia Fucus spiralis
6	Ascophyllum	Porphyra	Fucus vesiculosus Laurencia	Fucus spiralis
5	Ascophyllum Fucus vesiculosus	Porphyra Fucus spiralis	Fucus vesiculosus Laurencia	Ascophyllum Laurencia
4	Ascophyllum Fucus vesiculosus Fucus serratus	Porphyra Fucus spiralis Bare	Fucus vesiculosus Laurencia Fucus serratus	Ascophyllum Laurencia Fucus vesiculosus
3	Ascophyllum Fucus vesiculosus Fucus serratus	Fucus spiralis Bare	Laurencia Fucus serratus	Ascophyllum Fucus vesiculosus
2	Fucus vesiculosus Fucus serratus	Bare Fucus serratus	Laurencia Himanthalia	Ascophyllum
1	Fucus serratus	Laurencia Fucus serratus	Himanthalia	Fucus serratus
M.L.W.	Fucus serratus	Laminaria Laurencia Fucus serratus	—	Fucus serratus
−1	Fucus serratus	Laurencia Laminaria	—	Fucus serratus Laminaria
−2	Halidrys	—	Laminaria	Laminaria
−3	Halidrys Laminaria	—	Laminaria	Laminaria
Tidal range	8–9 ft.	6·5 ft.	18 ft.	10 ft.

zones are often 1 or 2 ft. higher than might otherwise have been
expected, and the height by which these zones are elevated is
termed the "splash zone". At Peveril Point the splash zone is
about 1 ft., whereas at Wembury in Dorset and Mount Desert
Island in Maine it is computed at 2 ft.

There are some features of particular interest from the individual localities which may suitably be discussed at this stage. Anand (1937) in his study of the Dover cliffs carried out some experiments with a view to determining the nature of the controlling factors. The greatest attention was given to the water relations, and it was pointed out that the water content of the algal covering depends on

(a) the supply of water, e.g. tides, spray, rainfall, humidity, moisture of the substrate;

(b) the water loss due to various causes, e.g. evaporation, drainage and capillary attraction of neighbouring belts, the latter being effective up to a distance of 40 cm.;

(c) the physical nature of the algal covering, e.g. whether delicate plants, leathery plants or gelatinous plants, whilst the quantity of water retained will also depend upon the thickness of the algal mat.

It was found that the *Enteromorpha* mat lost 25 % of its moisture in the first 3 hours of exposure, whilst the Chrysophyceae belt lost 18·4 %. The relative loss by evaporation of *Enteromorpha* and Chrysophyceae mats is seen in fig. 195 A, whilst the corresponding loss due to drainage is shown in fig. 195 B, the two sets of measurements being obtained by the simple but ingenious method of weighing portions of the mat cut out so that they fit into waterproof paper dishes which could be put back into position on the shore. The differences in loss for both evaporation and drainage are due to the gelatinous nature of the Chrysophyceae belt, and this result is obtained in spite of the fact that the evaporating power of the air opposite the latter belt is 1·41 as compared to 1·1 opposite the *Enteromorpha* belt, the evaporating power of the *Fucus* belt being taken as unity.

Similarly, the concentration of salt in the Chrysophyceae belt, which shows little variation, can be compared with that of the *Enteromorpha* belt which varies considerably with level and length of time after fall of the tide. The day temperature of the belts only responds to changes of air temperature in summer, and then it is always less than that of the air although the seasonal range is greater. The temperature range is greatest in the *Fucus* and least in the Chrysophyceae belt as the latter retains more moisture. If, however, the period of insolation is at all long, as may well happen in the summer months, then the Chrysophyceae mats frequently become cracked and fall off.

Light, currents and temperature are the chief factors determining the incidence of cave vegetation, winding caves showing the influence of light best. Lack of sunlight stops *Fucus* from invading these areas and when the light intensity is low the Chrysophyceae also are not able to develop satisfactorily. There is no algal growth

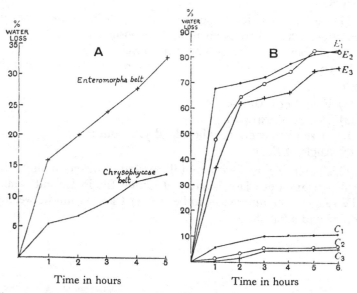

Fig. 193. A, water loss from samples of *Enteromorpha* and Chrysophyceae belts when exposed in their original position on the cliff face. B, water loss during drainage in nature from different levels in the *Enteromorpha* and Chrysophyceae belts during successive hours in winter. E_1–E_3, C_1–C_3 = successive levels. Water loss in A and B expressed as % of that originally present. (After Anand.)

in long and relatively straight caves beyond a distance of about 15·5 m. from the entrance where the light intensity has been reduced to about 1·8 % of the light outside.

A somewhat different ecological approach was adopted by Colman (1933) at Wembury. He carried out statistical analyses, and these showed that so far as the fauna and flora were concerned there are probably three critical levels:

(*a*) Between mean and extreme low-water marks of spring tides where the annual exposure is less than 5 %. This marks the lower limit of several intertidal species.

(b) Between the mean low-water marks of neap and spring tides where the exposure is about 20 %. This marks the upper limit of several submarine species.

(c) At the extreme high-water mark of neap tides where there is about 60 % exposure. This marks the upper limit of several inter-tidal species.

The least critical level appears to be mean low-water mark of neap tides where there is about 40 % exposure because the maximum number of species occurs at this level.

It can be seen therefore that the zonation depends very largely upon

(a) Extent of tidal rise.

(b) Degree of exposure.

To these two factors may be added yet a third:

(c) Angle of slope.

This latter feature is very well illustrated in the accounts of the algal vegetation at Clare Island and Lough Ine in Ireland. Table XIV sets out the differences to be seen at Clare Island between a sloping and a flat shore.

TABLE XIV

	Range	
	Sloping shore ft.	Flat shore yd.
Pelvetia zone	2–3	5
Fucus spiralis var. *platycarpus* zone	5–6	10
Ascophyllum zone	10	40
Ascophyllum and *F. vesiculosus* mixed	10	30
Fucus vesiculosus zone	10	50
Fucus serratus zone	10	50

A rather more detailed analysis of the same problem has been presented by Rees (1935) for Lough Ine.

Apart from the actual control of the zones themselves it has become increasingly evident that the position of a zone on the shore is to some extent determined by the temperatures of the different seasons. Attention was first drawn to this aspect of the problem by Knight and Parke (1931) in their work on the algal flora of the Isle of Man. They showed, for example, that successive generations of *Cladophora rupestris* move vertically up and down 5–8 ft. each year, the movement being rendered possible because

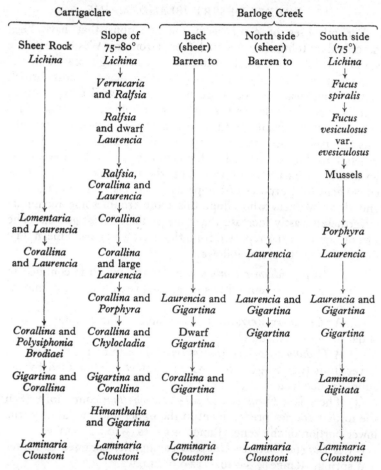

Carrigaclare		Barloge Creek		
Sheer Rock	Slope of 75–80°	Back (sheer)	North side (sheer)	South side (75°)
Lichina	*Lichina*	Barren to	Barren to	*Lichina*
	Verrucaria and *Ralfsia*			*Fucus spiralis*
	Ralfsia and dwarf *Laurencia*			*Fucus vesiculosus* var. *evesiculosus*
	Ralfsia, Corallina and *Laurencia*			Mussels
Lomentaria and *Laurencia*	*Corallina*			
				Porphyra
Corallina and *Laurencia*	*Corallina* and large *Laurencia*		*Laurencia*	*Laurencia*
	Corallina and *Porphyra*	*Laurencia* and *Gigartina*	*Laurencia* and *Gigartina*	*Laurencia* and *Gigartina*
Corallina and *Polysiphonia Brodiaei*	*Corallina* and *Chylocladia*	Dwarf *Gigartina*	*Gigartina*	*Gigartina*
Gigartina and *Corallina*	*Gigartina* and *Corallina*	*Corallina* and *Gigartina*		*Laminaria digitata*
	Himanthalia and *Gigartina*			
Laminaria Cloustoni	*Laminaria Cloustoni*	*Laminaria Cloustoni*	*Laminaria Cloustoni*	*Laminaria Cloustoni*

there is a monthly reproduction when the sporelings only survive in the most favourable zone for the particular time of the year, and this is not necessarily that in which the parent plants are growing. Some algae migrate up in winter and down in summer, whilst others move up in summer and down in winter. It is suggested that the nature of the response to temperature controls the movements of those plants that migrate down in winter, whilst response to strong insolation determines the behaviour of those that move down in summer.

MOUNT DESERT ISLAND, MAINE

For purposes of comparison the zonations that have been described by Johnson and Skutch (1928) from this Western Atlantic station are of considerable interest. It is a rough coast and the vegetation is not only much poorer than that of most British stations but also the zones are less distinct, whilst exposure to storms is responsible for a splash zone of about 2 ft. The plant aspect may vary from season to season and from year to year, but this fact has already become emphasized in describing the British vegetation. Johnson and Skutch recommended levels based on sea level as the best means of recording the belts because it is more convenient for purposes of comparison. This is undoubtedly true, and those workers who adopt this more troublesome technique nevertheless vastly increase the value of their investigations. The littoral communities reported from this area, which has a mean tidal range of 10·4 ft., are as follows:

(1) A *Porphyridium cruentum* community, which is confined to the spray zone, has four other species associated with the dominant alga.

(2) A *Calothrix-Verrucaria* community which ranges from +9·0 to −12·0 ft. M.L.W.

(3) A *Codiolum* society that appears in summer only with its lower limit (range +6·0 to −12·0 ft. M.L.W.) determined by the submergence factor.

(4) There is a *Fucus vesiculosus—Ascophyllum* community with the former species predominant in the upper and the latter in the lower portion of the zone. Range +5·0 to −12·0 ft. M.L.W.

(5) A *Bangia-Ulothrix-Urospora* community confined to winter and spring. Range +8·0 to −12·0 ft. M.L.W.

(6) An *Enteromorpha* community which is purely aestival. Range +3·0 to −10·0 ft. M.L.W.

(7) A *Porphyra* community on the steep slopes from +2·0 to −6·0 ft. M.L.W.

(8) A *Fucus furcatus* community in the more shaded parts of the area with a range of +2·0 to −6·0 ft. M.L.W.

(9) *Rhodymenia* community. Range +2·0 to −5·0 ft. M.L.W.

(10) In some years a *Spongomorpha arcta* society can be found. Range +2·0 to −5·0 ft. M.L.W.

(11) A *Spongomorpha spinescens* society appears during the summer months. Range +2·0 to −7·0 ft. M.L.W.

Only one association is recorded from the sublittoral, but this ought to be subdivided into three communities if it is to be compared with British coasts.

Sublittoral community

(12) An *Alaria-Halosaccion-Lithothamnion* community in which the red alga *Halosaccion* is most abundant on sloping rocks although in some parts of the coast it is replaced by *Chondrus*. Range −6·0 to +2·0 ft. M.L.W.

Local societies of *Saccorhiza* may occur between −1·0 and +1·0 ft. M.L.W. The kelps and *Halosaccion* are usually so dense that they prevent the downward migration of species from the littoral zones above, although where there is any available space such a migration will readily occur. This illustrates the effect of competition in determining zonation.

ZONATION IN WARM WATERS

So far we have only described the vegetation of temperate and cold waters. It is, however, very instructive to consider briefly the algal ecology of warm waters and observe how it differs from that of the colder waters. In the Mediterranean, for example, considerably more attention has to be paid to the sublittoral region, partly because of the extensive vegetation that persists in such a place, and partly because the small tidal rise, 20–30 cm., renders this region much more important. If an ecological survey were to be carried out in the Caribbean a similar state of affairs would be encountered. Here the very small tidal rise of less than a foot means that there is practically no intertidal vegetation, and indeed, zonation of algal belts is very rare although it may occasionally be encountered on beach rock. The algal vegetation of the Caribbean is almost wholly sublittoral, the associations being determined very largely by the type of substrate. The extent to which the sublittoral in the Mediterranean is of importance is illustrated in Table XV which is a summary of the various schemes that have been proposed for classifying the vegetation from this region.

In concluding this section a word may be said about the periodicity of the vegetation in the Mediterranean as compared with that of the English Channel. First of all there is the same pronounced difference in the floral aspects of the summer and winter months that has been observed on other coasts. Boreal Atlantic species such as *Ulothrix flacca*, *U. subflaccida*, *U. pseudoflacca*, *Bangia fusco-purpurea* and *Porphyra* spp. dominate the flora in winter, whilst in summer it is the tropical and subtropical species such as *Siphonocladus pusillus*, *Acetabularia mediterranea*, *Pseudobryopsis myura*, *Liagora viscida*, etc., which form the dominant species. In comparing the behaviour of the Mediterranean vegetation with that of the Boreal Atlantic one may distinguish several types of algal periodicity:

(1) Algae with a summer vegetation period in both the English Channel and the Mediterranean. These algae usually occur at a considerable depth where there is little or no temperature variation, e.g. *Sporochnus pedunculatus*, *Arthrocladia villosa*.

(2) Algae with a winter and spring vegetational period in both the Mediterranean and the English Channel, e.g. species of cold waters such as *Ulothrix flacca*.

(3) Algae appearing in the winter and spring in the Mediterranean but during the summer in the English Channel. For these algae it might be supposed that the temperatures of the winter and spring in the Mediterranean correspond more or less to the summer temperatures of the Channel, e.g. *Nemalion helminthoides*.

(4) Algae found during the summer months in the English Channel but persisting throughout the year in the Mediterranean, e.g. *Padina pavonia*. Their absence in the Channel at other times of the year may be associated with the low temperatures, and it ought to prove possible to ascertain the minimum temperature at which such algae will survive.

(5) Algae of spring and winter in the Mediterranean but persisting throughout the year in the Channel, e.g. *Porphyra umbilicalis*, *Callithamnion corymbosum*. This again is probably related to a temperature correlation, but in this case the algae concerned will not tolerate the high temperatures that are reached during the summer months in the Mediterranean. Comparisons of this

TABLE XV. *Concordance of the different systems of classification for the marine belts in the Mediterranean*

E. Forbes (1842)	Lorenz (1863)	Ardissone and Strafforello (1877)	Pruvot (1895)	Flahault (1901)	Seural (1924)	Feldmann (1937)
Mean low-water mark	I. Region supralittoral	Subzone I	Subterrestial zone / Superior horizon	Subterrestial zone { Superior horizon	Superior horizon	Supralittoral belt / Littoral belt
	II. Region of emergent littoral				Middle horizon	
Region I						
		Zone I				
Region II	III. Region of submerged littoral	Subzone II / Subzone III	Middle horizon	Inferior horizon }	Inferior horizon	Upper infra-littoral belt
Region III						
				(Marine littoral zone) / *(Inter co-tidal belt)*		
Region IV	Region IV	Zone II	Lower horizon	Marine sub-littoral zone	Littoral belt	Lower infra-littoral belt
Region V	{ Region V	} Zone III				
Region VI	{ Region VI					
Region VII						
			(Littoral region / Littoral zone)			
Lower limit of vegetation						
Region VIII			Coastal region	Abyssal zone	Coastal belt	Elittoral belt

nature are extremely valuable in helping us to understand something of the biological requirements of the species in question. It is also evident that they indicate some very profitable lines of investigation concerning the temperature relations of algae.

REFERENCES

England. ANAND, P. (1937). *J. Ecol.* **25**, 153, 344.

England. BAKER, S. M. (1909, 1910). *New Phytol.* **8**, 196; **9**, 54.

England. COLMAN, J. S. (1933). *J. Mar. Biol. Ass. U.K.* **18**, 435.

Ireland. COTTON, A. D. (1912). Clare Island Survey. Part XV. *Sci. Proc. R. Dubl. Soc.* **31**.

New Zealand. CRANWELL, L. M. and MOORE, L. B. (1938). *Trans. Roy. Soc. N.Z.* **67**, 375.

Mediterranean. FELDMANN, J. (1937). *Rev. Alg.* **10**, 1.

England. GIBB, D. (1938). *J. Ecol.* **26**, 96.

Scotland. GIBB, D. (1939). *J. Ecol.* **27**, 364.

England. GRUBB, V. M. (1936). *J. Ecol.* **24**, 392.

North America. JOHNSON, D. S. and SKUTCH, A. S. (1928). *Ecology*, **9**, 188.

England. KNIGHT, M. and PARKE, M. W. (1931). *Manx Algae*, p. 27. Liverpool.

Ireland. REES, T. K. (1935). *J. Ecol.* **23**, 69.

England. REES, T. K. (1940). *J. Ecol.* **28**, 403.

CHAPTER XII

ECOLOGY OF SALT MARSHES

In comparison with the rocky coast fewer studies have been carried out on the algal ecology of salt marshes, but those that have been published can be regarded as having made considerable advances in our knowledge of these extremely interesting areas. Their neglect in the past has probably been due to the fact that the algae are often microscopic and hence not so pleasing aesthetically even when present in abundance, and also they are more difficult to determine taxonomically. In practice, however, a detailed study of any one area often produces the rather unexpected result of a very extensive flora. For example, the number of species recorded from the English salt marshes of Norfolk is about two hundred, which does not compare unfavourably with the number on a rocky coast.

An investigation of any salt-marsh area shows that the algal communities offer a somewhat different aspect to the algal communities of a rocky coast. In the latter case it has been seen that zonation is a characteristic feature together with some super-imposed seasonal changes and migrations. On the salt marshes it is not really possible to distinguish any zonation but there may be well-marked seasonal changes in any one area. Thus on a fairly low marsh the "Autumn Cyanophyceae" appear in autumn and early winter, they disappear and are replaced in spring by the *Ulothrix* community, which in its turn is replaced during the summer months by *Enteromorpha* and so the cycle proceeds. Furthermore, as each year the ground level increases in height in relation to the tide through the continual deposition of silt, the submergences become fewer and the communities are replaced by others on account of the modified conditions. As a result there is a definite dynamic succession of the different communities over a long period of years. This cannot be seen on a rocky coast where there is no succession in time and where the succession in space is static.

The phenomenon of dynamic succession in this type of habitat necessitates a somewhat different approach to the problem of the status of the community. The continual replacement of one

C S A   21 community by another as the marsh increases in height provides changes that are more akin to those that are found in land habitats. With this in mind the present author recently attempted a survey of our present information about the algal communities of salt marshes. The principal features are set out in Table XVI, and it will be observed that in the suggested nomenclature the ordinary ecological terminations for developing seres has been employed. Whether this is entirely justified in view of the present somewhat scanty knowledge may perhaps be questioned, but it is possible that if the nomenclature can be placed on a proper basis at an early stage it should facilitate future comparisons.

Table XVI shows that there is not the same ubiquity of the communities in the different areas that can be found on a rocky coast. The reason for this is probably to be associated with the very different types of salt marsh that can be found. For example, the Irish marshes are composed of a form of marine peat, the marshes on the west coast of England have a large sand component in the soil, the marshes on the south coast bear a tall vegetation of *Spartina* growing in a very soft mud, whilst the east coast marshes bear a very mixed vegetation growing on a mud that tends to be clay-like. In spite of this, however, the Sandy Chlorophyceae, Muddy Chlorophyceae, Gelatinous Cyanophyceae, *Rivularia-Phaeococcus* socies, *Catenella-Bostrychia* consocies and the *Fucus limicola* consocies all have a wide distribution though they may not necessarily appear at the same relative levels on the different marshes. On the whole, however, they are very often found in the same phanerogamic community.

A comprehensive tour of the salt marshes of England will show us that one or more of the communities described above occur in all the different districts. Where the soil is rather sandy a *Vaucherietum* can be distinguished dominated by *V. sphaerospora*, but where the phanerogamic vegetation is very dense or heavily grazed by animals the algal vegetation is poor, e.g. south and west coast marshes. The Sandy Chlorophyceae and *Vaucheria Thuretii* have a wide distribution, as also the *Catenella-Bostrychia* community, whilst the pan flora appears to be richest in East Anglia. Perhaps the most interesting feature is the distribution of the marsh fucoid *Pelvetia canaliculata* ecad *libera* which occurs in north Norfolk, Lough Ine and Strangford Lough in Ireland and at Aberlady near

TABLE XVI. *Salt-marsh algal communities*

No.	Scolt Head Island	Canvey and Dovey	Clare Island	Lough Ine, Ireland	Communities, Suggested nomenclature
I	General Chlorophyceae	General Chlorophyceae	—	—	General Chlorophyceae associe
I a	Low sandy Chlorophyceae	—	—	—	Low sandy Chlorophyceae consocies
I b	Sandy Chlorophyceae	—	Sandy Chlorophyceae	Filamentous algae	Sandy Chlorophyceae consocies
I c	Muddy Chlorophyceae	—	Muddy Chlorophyceae	—	Muddy Chlorophyceae consocies
II	Probably present. Not investigated	Marginal diatoms	—	—	Marginal diatom consociation
III	Marginal Cyanophyceae	Marginal Cyanophyceae	?	Vertical banks ass.: (a) *Rhizoclonium-Phormidium* band (b) *Ulothrix-Phormidium* band	Marginal Cyanophyceae consociation
IV	*Ulothrix* community	*Ulothrix flacca* community	—	—	Vernal *Ulothrix* socies
V	*Enteromorpha minima* community	*Enteromorpha minima-Rhizoclonium* community	—	—	*Enteromorpha minima* socies
VI	Gelatinous Cyanophyceae	*Anabaena torulosa* community	—	Gelatinous Cyano-phyceae association	Gelatinous Cyanophyceae, socies or society depending on permanence
VII	Probably present. Not investigated	Filamentous diatoms	—	—	Filamentous diatom consocies
VIII	Autumn Cyanophyceae	Autumn Cyanophyceae	—	—	Autumn Cyanophyceae consocies
IX	*Phormidium autumnale* community	*Phormidium autumnale* community	—	—	*Phormidium autumnale* socies
X	*Rivularia-Phaeococcus* community	*Rivularia-Phaeococcus* community	*Rivularia-Phaeococcus* association	*Rivularia* association	*Rivularia-Phaeococcus* socies
XI	—	*Pelvetia muscoides* community	—	—	Dwarf *Pelvetia* socies
XII	*Catenella-Bostrychia* community	*Catenella-Bostrychia* community	*Catenella-Bostrychia* association	*Catenella-Bostrychia* association	*Catenella-Bostrychia* consocies
XIII	*Pelvetia-Bostrychia* community	—	—	?	*Pelvetia limicola* consocies
XIV	*Enteromorpha clathrata* community	—	—	—	*Enteromorpha clathrata* socies
XV	*Fucus limicola* community	—	*Fucus limicola* association	Limicolous Fucaceae association	*Fucus limicola* consocies
XVI	Pan community	—	—	Pan association	Pan association
XVII	—	—	—	—	*Vaucheria* consocies

the Firth of Forth but with no apparent intermediate stations. The normal form is present in other areas where there are marshes in the vicinity, e.g. the west coast marshes, but the marsh form does not appear to have developed. The evidence at present available would suggest that it has originated independently in the three areas, and in that case it can only be concluded that certain conditions must be fulfilled before the marsh form can develop from the normal species. This is a problem that is still awaiting solution.

One of the more interesting features of the algal vegetation of salt marshes is the occurrence of the marsh fucoids. These are peculiar forms which are either free-living on the marsh or else embedded in the mud, and they must all at one time have been derived from the normal attached form. Sometimes they bear a fairly close resemblance to the attached form but in other cases they have been very considerably modified, and it is only the existence of intermediate forms which enables us to indicate the normal type from which they came. East Anglia is essentially the home of the marsh fucoids, although Strangford Lough in Ireland is also extremely rich. In Norfolk, for example, considerable areas can be found occupied by *Pelvetia canaliculata* ecad *libera*, whilst the three marsh forms of *Fucus vesiculosus*, ecads *volubilis*, *caespitosus* and *muscoides* are also abundant, the last two being embedded in the soil.

Apart from these forms there are three other loose-lying marsh forms derived from *Fucus vesiculosus* but these are confined to the Baltic, e.g. ecads *nanus*, *subecostatus* and *filiformis*. A small crawling marsh form derived from *F. ceranoides* has been described from the Irish and Dovey marshes, and another larger free-living one from Strangford Lough in Ireland; like many others of this type it is profusely branched, fertile conceptacles are rare and, when present, are invariably male. *F. spiralis* vars. *nanus* and *lutarius* are other marsh derivatives, whilst *Pelvetia canaliculata* not only gives rise to ecad *libera* but also to a small embedded form, ecad *radicans*, which has been recorded from the Dovey marshes. There is also another form, ecad *coralloides*, which has been described from Blakeney and more recently from the Cumbrae marshes, but until more is known about this particular ecological form it ought to be regarded with some degree of caution. *Ascophyllum nodosum* var. *minor* is a dwarf embedded variety, ecad *Mackaii* of the same species is a free-living form found on American salt marshes, in Scotland and on the shores of Strangford Lough in Ireland, whilst ecad *scorpioides* is a partially embedded form found on the Essex marshes and on the shores of Strangford Lough. All these forms probably originated as a result of vegetative budding, although it is also possible that they have developed from fertilized oogonia that became attached to phanerogams on the marsh. There is definite evidence that *Ascophyllum nodosum* ecad *scorpioides* arises by vegetative budding from fragments of the normal plant, whilst it has been suggested that conditions of darkness may be favourable for the development of ecad *Mackaii*. As a group the marsh fucoids are characterized by

(1) vegetative reproduction as the common means of perpetuation;

(2) absence of any definite attachment disk;

(3) dwarf habit;

(4) curling or spirality of thallus.

In the embedded forms derived from *Fucus vesiculosus* the three-sided juvenile condition (cf. p. 197) of the apical cell is retained, the cryptostomata are marginal and division in the megasporangia is only partial or else does not occur. It is suggested that these features are due to

(*a*) exposure, which results in a dwarfing of the thallus;

(*b*) lack of nutrient salts which induces a narrow thallus;

(*c*) the procumbent habit and consequent contact with the soil causes spirality because growth takes place more rapidly on the side touching the soil.

The cause of sterility may either be a result of the high humidity (according to Baker, 1912, 1915) or, more probably, because of the persistence of the juvenile condition as represented by the apical cell and cryptostomata. The marsh fucoids occur most frequently either as pioneers on the lowest marshes or else as an undergrowth to the phanerogams.

One of the more striking physiographical features of salt marshes is the salt pan. The number, shape and size of these on the different salt marshes varies very considerably, but they generally contain a certain number of algae, especially those pans which occur on the lower marshes. They are important because they provide a much wetter habitat at levels where normally conditions may be somewhat dry. Some authors are not prepared to acknowledge the existence of a pan flora because they maintain that the plants are not persistent. A continual study of pans in one area over a considerable period of time by the present author showed that a definite pan flora did exist from year to year, and that many of the species comprising it reproduce during the course of their existence. The mere fact that they can carry out normal reproduction would seem to validate the recognition of such a flora.

There are several interesting features concerning the pan flora of the Norfolk salt marshes which may conveniently be mentioned here. There are two different types of salt pan, those with soft floors and those with firm, the algal flora usually being confined to the latter, although so far there is no explanation of this feature. On the lower marshes the pan flora is commonly composed of Chlorophyceae, whilst with increasing marsh height the Chlorophycean element decreases and the Cyanophycean element increases. A few of the constituent members, e.g. *Monostroma*, are seasonal in appearance, whilst on some marshes there are pans which contain algae that are normally associated with a rocky shore, e.g. *Colpomenia, Polysiphonia, Striaria*. These persist from year to year in spite of the stagnant conditions, and when compared with the habitats occupied by the same species on a rocky coast it is found that they are probably growing at an unusually high level. Comparing the Norfolk marsh flora with that of a rocky coast the following two generalizations can be made:

(*a*) Species that are littoral on a rocky coast are to be found growing at lower levels, usually sublittoral, on the marsh coast. This must be ascribed to the lack of a solid substrate at the higher levels where they would normally grow.

(*b*) Littoral species of the rocky coast are found growing at higher levels on the marsh coast. This can be understood in the case of those species living in pans or in the streams where they are continually covered by water, but at present it is difficult to provide an explanation for the few species which actually grow on the marshes.

Turning now to the algal vegetation of the marshes proper, Carter (1932, 1933) has suggested that on the Canvey and Dovey marshes light and space relations, rather than factors relating to level, influence the distribution of the various species. Whilst this is undoubtedly true there is no doubt that the increasing height of a marsh with its consequent greater exposure does nevertheless effectively determine the upper height to which many plants can go. The species to be found on the higher marshes in Norfolk are either fucoids or gelatinous Cyanophyceae, both of which have the power of retaining moisture. The more delicate Chlorophyceae are more or less confined to the lower levels. On the other hand a

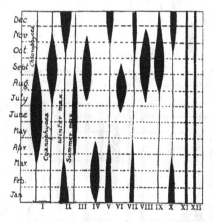

Fig. 194. Distribution in time of the algal communities on the salt marshes at Canvey and Dovey. I. General Chlorophyceae. II. Marginal diatoms (two components, (A) those with a winter maximum; (B) those with a summer maximum). III. Marginal Cyanophyceae. IV. *Ulothrix* community. V. *Enteromorpha minima*. VI. *Anabaena torulosa*. VII. Filamentous diatoms. VIII. Autumn Cyanophyceae. IX. *Phormidium autumnale*. X. *Rivularia-Phaeococcus*. XI. *Pelvetia canaliculata*. XII. *Catenella-Bostrychia*. (After Carter.)

dense phanerogamic vegetation, such as one finds on the south coast marshes where the tall *Spartina Townsendii* must lower the light intensity considerably, does reduce the quantity of algal vegetation. A similar state of affairs has been observed on the grass-covered marshes of New England.

From data available it is possible to compare the distribution in space (e.g. among the different phanerogamic communities) and time of the marsh communities recorded from Canvey, Dovey and Norfolk. Figs. 194 and 195 show the distribution of the Canvey and Dovey communities and they should be compared with figs. 196

and 197 for similar marsh communities of Norfolk. Some of the
smaller communities, e.g. *Rivularia-Phaeococcus* and Gelatinous

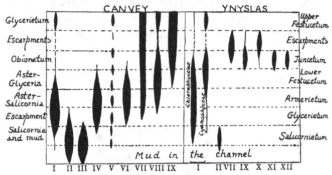

Fig. 195. Distribution of the algal communities in space on the Canvey and
Dovey marshes. Symbols as in Fig. 194. (After Carter.)

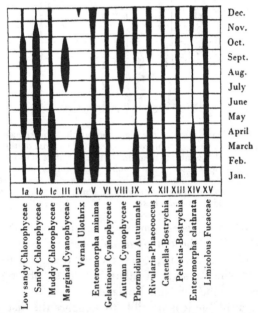

Fig. 196. Distribution of the algal communities throughout the year at Scolt,
Norfolk. (After Chapman.)

Cyanophyceae, are apt to be overlooked in summer because the
constituent species shrivel up so much or else because the colonies

become covered by an efflorescence of salt. An examination of the distribution of the various communities on the Norfolk marshes shows that five communities are each confined to one type of habitat. This relationship may be due to:

(*a*) Association with a particular phanerogamic community, e.g. *Phormidium autumnale* (IX) and *Obione portulacoides*.

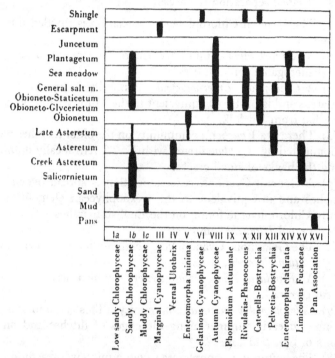

Fig. 197. Distribution of the algal communities in space at Scolt, Norfolk. (After Chapman.)

(*b*) Dependence upon certain edaphic conditions, e.g. Muddy Chlorophyceae (I*c*).

(*c*) Dependence upon the physical character of the environment, for example slope, exposure, wave action, e.g. Marginal Cyanophyceae (III), Vernal *Ulothrix* (IV) and the Pan Association (XVI).

MARSHES OF NEW ENGLAND

A study of one of the New England marshes near Boston by the present author has revealed the existence of the following algal communities which should be compared with those in Table XVI. It will be seen that there is considerable similarity in spite of the great distance separating the areas:

(1) The General Chlorophyceae association is divided into two components.

(a) A *Rhizoclonium* community occurs widespread in nearly all the phaneroganic communities.

(b) A *Cladophora-Enteromorpha* community is present in very wet areas and is probably equivalent to the Muddy Chlorophyceae recorded from Norfolk.

(2) There is a *Vaucheria* community on the older marshes which is dominated by *V. sphaerospora* with *V. Thuretii* locally dominant along the banks of small creeks.

(3) A General Cyanophyceae association is spread over all the marshes and is equivalent to the Cyanophycean element in the Sandy Chlorophyceae found on the English marshes.

(4) Vernal *Ulothrix* community.

(5) The Gelatinous Cyanophyceae community is associated with the *Juncetum Gerardii*.

(6) A *Rivularia-Phaeococcus* society is also associated with *Juncus Gerardii*.

(7) *Enteromorpha minima* community. This is abundant in spring and early summer along the edges of ditches and on old plants of *Spartina* spp.

(8) The autumn Cyanophyceae community is present on the higher marshes and also in the salt pans.

(9) The Limicolous Fucaceae community is dominated by one of the following, *Ascophyllum nodosum* ecad *Mackaii*, *Fucus vesiculosus* ecad *volubilis* or *F. spiralis* ecad *lutarius*.

(10) Pan Flora.

An analysis of the tidal factors operating on the salt marshes of Norfolk (England), Lynn (Mass.) and Cold Spring Harbor (L. I.), has suggested that for some of the species common to the two areas the controlling factors must be the same.

REFERENCES

England. BAKER, S. M. and BLANDFORD, M. (1912, 1915). *J. Linn. Soc. (Bot.)*, **40**, 275; **43**, 325.
England. CARTER, N. (1932, 1933). *J. Ecol.* **20**, 341; **21**, 128, 385.
England. CHAPMAN, V. J. (1937). *J. Linn. Soc. (Bot.)*, **51**, 205.
England. CHAPMAN, V. J. (1939). *J. Ecol.* **27**, 160.
America. CHAPMAN, V. J. (1940). *J. Ecol.* **28**, 118.
Ireland. COTTON, A. D. (1912). Clare Island Survey, Part XV. *Sci. Proc. R. Dublin Soc.* **31**.
Ireland. REES, T. K. (1935). *J. Ecol.* **23**, 69.

CHAPTER XIII

FRESH-WATER ECOLOGY

One of the major problems in this branch of algal ecology appears to be the establishment of a successful classification upon which field studies can be based. Up to 1931 the outline given by West in 1916 was in current use but since then a scheme proposed by Fritsch (1931) has more or less taken its place. It would seem, however, that neither scheme alone is wholly satisfactory, but that a combination of the two provides a very suitable basis for workers in this field. An outline of such a combination of the two schemes is briefly described below.

A. Subaerial association.

This develops at its best in the tropics although it can also be found in temperate regions. In the latter, Protococcales and *Trentepohlia* form the principal elements, whilst in the tropics the Cyanophyceae and other members of the Trentepohleaceae (*Cephaleuros, Phycopeltis*, etc.) form the dominant components.

B. Association of dripping rocks.

This can be subdivided into

(i) Permanently attached communities:
    (*a*) On living material.
    (*b*) On dead organic material.
    (*c*) On the hard rock (Epilithic).
(ii) Temporarily attached communities:
    (*a*) On living material.
    (*b*) On dead organic material.
    (*c*) On the hard rock (Epilithic).

C. Aquatic associations.

These vary from season to season and frequently have a marked periodicity which is controlled by diverse factors. Four sub-

divisions of the aquatic associations can be recognized (see below)
and *each one* of the subdivisions can be treated as follows:

(1) Attached communities (frequently termed the *Benthos*):

A	B
Permanently attached	Temporarily attached

    (*a*) On living material.
    (*b*) On dead organic material.
    (*c*) On inorganic material (Epilithic).
    (*d*) In the silt.

(2) Floating macro-communities (Pleuston):

    (*a*) Originating from loose bottom forms.
    (*b*) Originating from epiphytic forms.
    (*c*) Wholly floating throughout.

(3) Loose-lying communities of the bottom.

(4) Plankton or floating communities:

    (i) Limnoplankton of lakes.
    (ii) Potamoplankton of slow rivers.
    (iii) Cryoplankton of the eternal snows.

In this last category we have red snow due to the presence of
Chlamydomonas nivalis; yellow snow with a flora of about twelve
species all containing much fat; green snow, principally caused by
the zoogonidia of green algae; brown snow due to the presence of
Mesotaenium and mineral matter; black snow caused by *Scotiella
nivalis* and *Rhaphidonema brevirostre*; and a light brownish purple
ice-bloom caused by a species of *Ancyclonema*.

West (1916) divided the Aquatic Associations into the following
four major subdivisions, each of which can be further subdivided
in the manner illustrated above:

(1) *Associations of rivers, rapids, and waterfalls.*

This is mainly composed of fresh-water Rhodophyceae, *Cladophora* spp., *Vaucheria* spp., Cyanophyceae and diatoms, whilst the
flora of hot springs may also be included here. Cyanophyceae are
usually the only constituents, the various species being capable of
secreting carbonate of lime or silica to form rock masses such as
travertine and sinter, the rate of deposition sometimes being as much
as 1·25–1·5 mm. in three days. The highest temperature recorded

for hot springs that contain living plants (*Phormidium laminosum*) is 87·5° C.

In slow-moving rivers there is a definite *Potamoplankton* divided into:

(*a*) *Eupotamic*, thriving in the stream and its backwaters.
(*b*) *Tychopotamic*, thriving only in the backwaters.
(*c*) *Autopotamic*, thriving only in the stream.

(2) *Associations of bogs and swamps.*

These are very mixed associations with little or no periodicity, probably because of the relatively uniform conditions. Zygnemaceae, desmids and diatoms are most frequent, the desmid element changing considerably with altitude and type of substrate, whilst the presence of *Utricularia* apparently also increases the number and variety of the desmid species.

(3) *Associations of ponds and ditches.*

The flora exists under very varied conditions with a regular or irregular periodicity. In the temperate regions Protococcales, Zygnemaceae (dominant in spring) and diatoms (dominant in winter) form the chief elements. There is usually not enough aeration to permit the larger filamentous forms to be present, and for this reason the ponds and ditches can be divided into:

(*a*) those containing Cladophoraceae, which suggests that the aeration is good;
(*b*) those without Cladophoraceae. The substrate and fauna are also important factors in determining the type of vegetation to be encountered. The flora of tropical ponds contrasts sharply with that of temperate regions for there is

(1) an excess of Cyanophyceae;
(2) the poor aeration results in a relative scarcity of *Cladophora* and *Rhizoclonium* together with the epiphytes associated with them, and their place is taken by *Pithophora*;
(3) a scarcity of *Vaucheria, Oedogonium*, Xanthophyceae and Ulotrichales;
(4) an abundance of filamentous desmids together with *Spirogyra*.

In America Transeau (1913) concluded that fresh-water pond algae can be divided into seven classes based on abundance, duration and reproductive season, these classes and their periodicity being represented in fig. 198.

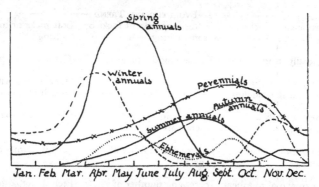

Fig. 198. Chart showing the estimated relative importance of the different types of algal periodicity throughout the year in the waters of E. Illinois. The irregulars are not depicted. (After Transeau from West.)

(4) *Associations of pools and lakes.*

West was the first investigator of lake and pool algae who appreciated the fact that the geology of the substrate was of profound importance. He showed that the desmid flora is richest where the substrate is precarboniferous, whilst diatoms become abundant in younger areas or where there has been much silting with consequent solution of mineral salts. Later workers have greatly extended this important study, and the present treatment of the problem is more or less summarized in the schema on p. 336.

A third type is the *Dystrophic* lake or pool, which is to be found on moorlands, where desmids form the most abundant part of the flora in a water that is often highly coloured. In the course of years Oligotrophic waters may also change into Dystrophic waters. In sheltered lakes as compared with open lakes there is an oxygen stratification which closely follows the bottom contours, whilst the influence of any rivers entering the lake together with the problem of periodic floods is yet a further factor.

Where there is a shallow littoral shore the communities are difficult to recognize unless there is a rocky substrate, in which case there may then be a zonation that is dependent on changes of water level and wave action: this type of zonation has been observed in several continental lakes. In deeper waters the communities are more distinct because a zonation develops which is primarily maintained by the light intensity factor. The Limno-

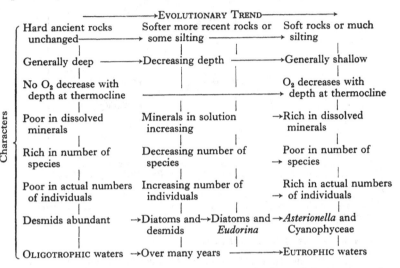

————————→EVOLUTIONARY TREND————————→

Characters			
Hard ancient rocks unchanged——→	Softer more recent rocks or some silting ——→		Soft rocks or much silting
Generally deep ——→	Decreasing depth ——→		Generally shallow
No O₂ decrease with depth at thermocline	——————→		O₂ decreases with depth at thermocline
Poor in dissolved minerals	Minerals in solution increasing		→Rich in dissolved minerals
Rich in number of species	Decreasing number of species		Poor in number of → species
Poor in actual numbers of individuals	Increasing number of individuals		Rich in actual numbers → of individuals
Desmids abundant	→Diatoms and desmids	→Diatoms and Eudorina	→*Asterionella* and Cyanophyceae
OLIGOTROPHIC waters	→Over many years ——————→		EUTROPHIC waters

plankton of lakes is not usually of great bulk and is composed principally of various members of the Cyanophyceae, Dinophyceae, Bacillariophyceae and Chlorophyceae, and according to the nature of the constituents it may exhibit maxima in spring (very commonly), spring and autumn or summer and autumn.

ASSOCIATIONS OF RIVERS AND STREAMS

Budde (1928) investigated very thoroughly the mountain streams feeding the Ruhr river. Most of these streams and brooks are trout streams and they can be divided into two regions:

(1) upper *Hildenbrandtia* region dominated by *H. rivularis*,
(2) lower *Lemanea* region dominated by *L. fluviatilis*.

The seasonal facies were studied and were found to be as follows:

(*a*) Spring period characterized by the dominance of diatoms with *Ulothrix* and *Hormidium* as subdominants.

(*b*) Summer period with Chlorophyceae and desmids predominant.

(c) Winter period during which *Ulothrix* and *Hormidium* reappear and the Diatomaceae increase.

The most important controlling factor is apparently temperature whilst the chemistry of the water is also significant, although local modifications of the flora may be brought about by changes of light intensity and oxygen concentration. When compared with the floras of streams from other areas it is interesting to note that the same species often occur in widely different types of habitat, thus providing a proof of the indifference of those plants towards habitat. The algal communities could be divided into three groups, those occupying a vertical substrate, e.g. waterfalls, those occurring on a horizontal substrate and those which are free-living.

A. Algal communities of vertical substrates:

(1) Those attached to stones; eight communities were distinguished.

(2) Epiphytic communities; four were distinguished, three of which also occur in (1).

(3) Three spray communities.

B. Algal communities of horizontal substrates:

(1) Those attached to stones, sand or mud; three communities.

(2) Six epiphytic communities.

C. Free-living algal communities: two were distinguished.

In a study of the encrusting algae of streams Fritsch (1929) distinguished (a) filamentous algae, (b) algae of banks and (c) submerged encrusting algae: in the particular stream there appeared to be a brief succession terminating in a mat of *Phormidium autumnale*.

Interesting results have also been obtained from a study of colonies of the blue-green alga, *Rivularia haematites*, growing in a stream. It was found that the surface area of the thallus increases greatly in proportion to the attachment area until finally the force of the torrent becomes greater than the prehensile force and the thallus is torn away. In fast streams the thalli are only formed on big stones because the small stones together with the colonies have been swept away. In such fast-flowing regions there appears to be a relationship between size of thallus and size of stone, but no such correlation can be demonstrated in quiet waters.

These facts are important because they serve to indicate that purely mechanical factors may be concerned in the distribution of some algae, and only too often this aspect of an ecological problem is wholly neglected.

POND ASSOCIATIONS

The literature on this subject is relatively sparse, but it is evident that periodicity of the different species is of paramount importance, the appearance of the different plants being controlled by a series of factors, only one of which may be limiting for any given species. A study of a pond near Harpenden by Fritsch and Rich (1913) showed that the general aspect of the flora was dependent upon season and that four phases could be distinguished:

(*a*) Winter phase with *Microspora*, *Eunotia* and epiphytic diatoms, whilst *Ranunculus aquatilis* and *Callitriche* were the dominant phanerogams.

(*b*) Spring phase dominated by Conjugatae, *Oedogonium* and *Conferva*, with *Ranunculus aquatilis* as the most important phanerogam.

(*c*) Summer phase with *Euglena*, desmids and *Anabaena* associated with a phanerogamic vegetation of *Lemna*, *Glyceria* and *Bidens*.

(*d*) Sparse autumn phase with *Lyngbya* and *Trachelomonas* but without any dominant phanerogam.

The algal periodicity is thus more or less associated with a similar periodicity in the phanerogamic vegetation. The flora differs from that of a similar pool near Bristol in the absence of *Cladophora* and *Melosira*, and in their place there is a greater development of Xanthophyceae. The two types of flora could be regarded as distinct associations, but the difference is almost certainly due to poor aeration in the Harpenden pool. In spite of this the general trend of periodicity in the two pools is very similar: a winter phase characterized by a hardy filamentous form (*Cladophora* or *Microspora*) and diatoms, a spring phase with Zygnemaceae and an autumn phase with Oscillatoriaceae. The summer phase in the two pools is very different, and this is ascribed to the greater drying up of the Harpenden pool during that period. The flora of pools, therefore, is very dependent not only upon general climatic conditions, such as rainfall and insolation, but also upon what might be termed irregular microclimatic factors, e.g. aeration in the body of water itself. In the case of many of the species there is a profound relationship between the meteorological data and the frequency of the flora, e.g. *Microspora* and the Protococcales with temperature, *Oedogonium* and *Hormidium* with sunshine. The factors influencing the growth of aquatic algae are (1) seasonal, (2) irregular, (3) correlated. The first group, which are very obvious and need not be detailed, are principally of importance for large bodies of water, but they tend to be masked by the other two groups in small bodies of water:

(2) *Irregular factors.*

    (1) Abnormal rainfall:

        (*a*) Species favoured by excessive rainfall.

        (*b*) Species favoured by drought.

    (2) Abnormal sunshine:

        (*a*) Species favoured by excessive sunshine.

        (*b*) Species adversely affected by excessive sunshine.

    (3) Abnormal temperature:

        (*a*) Species favoured by low temperatures.

        (*b*) Species favoured by relatively low temperatures.

        (*c*) Species favoured by high temperatures.

(3) *Correlated factors.*

    (1) Species depending on the enrichment of the water by decay of other members of the flora.

    (2) Forms influenced in their development by competition with others.

    (3) Forms influenced in their development by the presence of a suitable host, e.g. epiphytic forms.

A very definite correlation can frequently be established between the amount of sunshine and the phenomenon of reproduction, the latter process being most frequent when there is most sunshine. This is in accordance with experimental work by Klebs (1896) who showed that reproduction was initiated by the presence of bright light. An unusual concentration of the salts in the water during a period of drought may, however, counteract the influence of sun-shine.

A study of algal periodicity in some ponds near Sheffield, together with the results of fortnightly analyses, has suggested a correlation with the nitrate factor for some species. The maximum for this occurs in December whilst there is a minimum in June, and it was observed that *Volvox* received a severe check when the nitrate was high and only reproduced at times of low nitrate value. *Ulothrix* reappeared yearly in these ponds, whilst *Euglena* annually

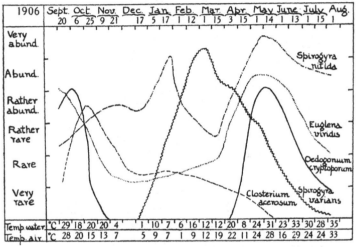

Fig. 199. Abundance and frequencies of the most important algae in a pond near Indiana University from 1906 to 1907. (After Brown.)

attained to a maximum between July and August soon after the nitrate minimum.

A similar study by Brown (1908) of some pools near the University of Indiana, revealed the fact that the species tended to attain their maximum abundance in autumn and spring. In one pond (fig. 199) the phases were as follows:

Phase	Dominants
Autumn	*Closterium, Euglena, Oedogonium*
Winter	*Spirogyra* spp.
Early spring	*Spirogyra* sp.
Late spring	*Spirogyra, Euglena, Oedogonium*
Summer	No one species

In another pond somewhat different phases were recorded:

Phase	Dominants
Autumn	*Oedogonium, Chaetophora*
Winter	*Vaucheria*
Late spring	*Oedogonium, Protococcus*
Summer	*Chaetophora*

These observations should be compared with those from the Harpenden pool, and it will be seen that although the spring phases are essentially similar with either *Spirogyra* or *Oedogonium*, nevertheless there are great differences. The two ponds described above also possessed floras that were essentially different and they must therefore be regarded as containing two separate associations. Furthermore, the same worker found that a sudden change in the external conditions checked the growth of an alga and often resulted in the development of a resting stage or else of sexual organs; insistence upon the importance of external conditions in this respect has also been emphasized by Fritsch and Rich in their study on the Harpenden and Bristol pools.

LAKE ASSOCIATIONS

Only one example of the algal flora of lakes will be discussed in these pages, and so the student must remember that lakes from other parts of the world may exhibit differences not only in species but also in the periodicity of the communities. A recent study by Godward in 1937 of the littoral algal flora of Windermere in Cumberland brought out a number of interesting facts. In the continental lakes, some of which are of a considerable depth, many of the algal communities are markedly limited in the depth to which they can descend. In Windermere, however, any species of the deeper waters is also able to exist in the surface layers, but as only a shallow depth of water is occupied by the various communities, depth *per se* can only be employed on a broad basis as a means of distinguishing the communities.

Three different groups of communities were recognized:

(1) Communities growing on stones and rocks:

 (a) Spray zone dominated by *Pleurocapsa* (May–September), *Tolypothrix* and *Phormidium* (April–September).

 (b) Zone 0–0·5 m. Dominated by *Ulothrix*, diatoms and Cyanophyceae.

(c) Zone 0–3·5 m. No definite community is formed in this belt.

(d) Zone 2–3·5 m. A distinct community dominated by Cyanophyceae.

(2) Epiphytic communities growing on aquatic macrophytes:

(a) On submerged plants between 0 and 0·5 m. This possesses a conspicuous Chlorophycean element, e.g. Conjugales, Chaetophorales and Ulothricales.

(b) On submerged plants between 1 and 3 m. dominated by *Oedogonium*, *Coleochaete* and diatoms.

(c) A community on submerged plants between 3 and 6 m. which is comprised of *Coleochaete*, a few diatoms and some Cyanophyceae.

(3) Communities on dead leaves and organic débris:

(a) Between 0 and 12 m.: wholly Diatomaceae.

(b) Between 2 and 16 m.: four diatom species and *Microcoleus delicatulus*.

The depth range of the diatoms was found to be greatest at the time of their maximum in spring and smallest in mid-winter. It was also discovered that the diatom frequency and light intensity often show an opposite trend in the upper layers and a similar trend in the lower layers of the lake. The nature of the habitat, whether organic or inorganic, makes a considerable difference to the behaviour of the different species, and each individual species responds to the differences of these two environments in its own way. In spite of these differences, however, they all exhibit an April maximum and depth has the same influence on them all (cf. fig. 200). A study of the plankton of Lake Windermere gave results that were in accordance with the view that the constituents of the floating community originate from the algae of the littoral region.

The periodical development of the littoral algal flora can be summarized as in Table XVII.

A study of the chemistry of the waters in the different algal habitats around the lake is summarized in Table XVIII.

An investigation of the distribution of the algae in relation to the different habitats showed that the algal species clearly fall into two main groups.

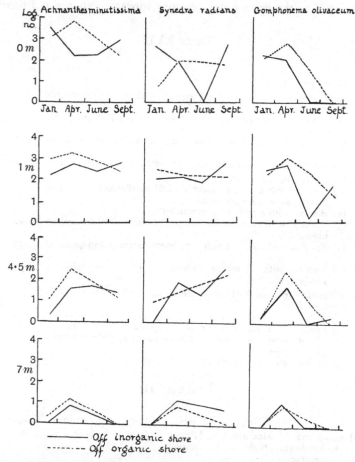

Fig. 200. Distribution of diatoms on slides suspended at different depths at different seasons of the year off two types of shore. (After Godward.)

A. Those typical of the inner parts of reed swamps (organic shores).
B. Those typical of other habitats:
    (i) Species more abundant in streams.
    (ii) Species more abundant on inorganic stony shores.
    (iii) Species more abundant in the outer parts of highly evolved reed-swamp and throughout the less evolved reed swamp.

A very definite gradation or succession can be traced in the algal flora as one passes from the inner to the outer reed swamps, from the latter to the open water or stony inorganic shores and finally

TABLE XVII

	No. of species
A. Occurrence of species	
(1) Species present throughout the year with no distinguishable maximum	4
(2) Species present throughout the year with a maximum at one period	4
(3) Species present in abundance only at certain times of the year	7
(4) Species present in some degree at certain times of the year	Numerous
B. Occurrence of maximum	
(1) Species with a spring maximum and smaller autumn maximum; diatoms predominant	9
(2) Species with a spring maximum only	3
(3) Species with a summer maximum only; Chlorophyceae predominate	11
(4) Species with an autumn maximum only; Cyanophyceae predominate	11
(5) Species with a winter maximum only; Chlorophyceae predominate	7
C. Time of year when different species occur in abundance at their greatest depth	
(1) Species attaining greatest depth in spring; diatoms only	3
(2) Species attaining greatest depth in spring and autumn	3
(3) Species attaining greatest depth in summer; Chlorophyceae predominate	10
(4) Species attaining greatest depth in autumn	1

TABLE XVIII

	NO_3	NH_3	P_2O_5	Organic matter	CO_2
(a) Stony and rocky shores (inorganic)	Moderately high	Low or absent	Low	Low	Low
(b) Mouths of streams	High	High	High	Variable	High
(c) Reed swamps (organic)	Low	High	Low	High	Very high

to the mouth of streams. In other words a progressive change in the algal flora is associated with a bottom that becomes less and less organic in nature or as one passes from eutrophic to oligotrophic conditions.

EPIPHYTES

It is convenient at this point to consider what is known about the distribution of algal epiphytes, and in this connexion a study of two ponds on the outskirts of Epping Forest by Godward (1934) has resulted in considerable advances to our knowledge. Three series of epiphytes were distinguished.

(1) Winter forms. 16 species approx.

(2) Summer and autumn forms. 11 species approx.

(3) Forms existing throughout the year. 11 species approx.

An investigation of the effect of the age of the substrate upon the epiphytic flora showed that the nature of the substrate was of great importance. This is illustrated in fig. 201 E, where it can be seen that, so far as the tips of the leaves are concerned, the total number of epiphytes increases up to the third or fourth leaf from the apex, after which there is a decline. The diatom flora, however, is an exception to this behaviour, because they increase regularly with the age of the substrate so that the oldest leaves bear the greatest number of diatomaceous epiphytes. On the other hand, algal zoospores tend to settle on the younger living leaves. There are distinct differences in the epiphytic flora of the upper and lower surfaces of leaves, and it was observed that in the case of the first few leaves below the apex the upper surface was infinitely superior, probably because of the greater light intensity. In addition to distribution·in relation to increasing age, there is also the relation to the different parts of the phanerogamic substrate. Fig. 201 E illustrates the distribution of epiphytes on the different parts of a phanerogam, and it will be observed that it is only on the leaf tips that the maximum is reached at the third or fourth leaf whilst the leaf sheaths show a slight maximum at about the tenth leaf with a well-marked maximum for the mid-rib at the same level. These maxima on the lower leaves are to be associated with the diatom flora. It will also be observed that the number of epiphytes on the internodes remains more or less constant, but rapid growth of the substrate, e.g. the leaf lamina, tends to prevent colonization by epiphytes. The density of epiphytes that are attached to dead organic material is dependent upon the habitat of the substrate, e.g. if it is floating then there are few epiphytes, if it is attached or submerged the epiphytes are numerous, whilst if it is lying on the bottom the epiphytes will be few. The various species to be found are all a residuum from the last living state of the material, and the assertion that dead material bears more epiphytes than living does not appear to be correct in this case and it can only be supposed that it arose in the past through lack of quantitative analysis. In some cases the appearance of epiphytes is due to change in the host with age, e.g. old filaments of the Zygnemaceae lose their mucilage sheath and they then become colonized by many epiphytes.

Experimental work and observation show that the greatest growth and number of epiphytes are partly related to conditions of good illumination, a feature which is illustrated by Table XIX below.

TABLE XIX

Level	Total no. of epiphytes collected on suspended slides	
	Sandy bottom	Muddy bottom
Water level	225 (no *Eunotia*)	262 (no *Eunotia*)
5 cm.	176 (no *Eunotia*)	108 (102 *Eunotia*)
12 cm.	176 (no *Eunotia*)	0
17 cm.	37 (all *Eunotia*)	0

When considering the effect of illumination it has to be remembered that not only are there problems associated with the individual plants, such as the upper and lower surfaces of leaves, but also that the density of the host plants may be highly significant. Fig. 201 A–D shows the distribution of various epiphytes on plants of *Equisetum limosum* under different conditions of spacing and the contrast is exceedingly obvious. Where there is screening of leaves, either on the same plant or by several plants, then the epiphytes develop on the unscreened portion.

The interrelations of host and epiphyte are important, and it was noticed that the epiphytes tend to develop in the depressions where the cells of the host adjoined each other. Experiments were then carried out with scratched slides suspended in the water, and the results obtained from these rendered it clear that depressions in a surface increase the number of epiphytes very considerably.

So far as the attachment organs of the epiphytes are concerned there is no apparent relation between the nature of the substrate and the method of attachment. The differences seen above, there-fore, must be explained by the behaviour of the motile reproductive bodies which either come to rest in the depressions or else are swept there by micro-currents in the water. Another interesting

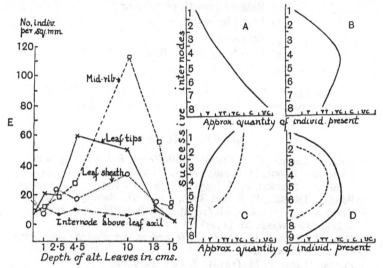

Fig. 201. A, B, distribution of *Cocconeis placentula* on successive internodes of plants of *Equisetum limosum*, well separated (3 stems average). r = less than 5 individuals per 0·1 sq. mm.; rr = about 5; rc = about 10; c = about 30; vc = about 50. C, distribution of *Cocconeis placentula* (——) and *Eunotia pectinata* (......) on crowded plants of *Equisetum limosum* (3 stems average). D, distribution of *Stigeoclonium* sp. (——) and *Coleochaete scutata* (......) on fairly crowded stems of *Equisetum limosum* (2 stems average). E, distribution of total epiphytes on successive leaves of *Oenanthe fluviatilis*. (After Godward.)

TABLE XX

Epiphyte	No. in scratches	No. elsewhere
Cocconeis	517	297
Stigeoclonium sp.	665	198
Chaetopeltis	138	54
Ulvella	747	200
Coleochaete scutata	40	13

feature is the frequent association of *Gomphonema* with the basal cells of *Oedogonium*, but so far there is no evidence to suggest whether this is a casual relationship or not. Ponds with muddy bottoms have a reduced number of epiphytes probably because the pH and the gases evolved are toxic, but so far little or no work has been carried out to ascertain the effect of the host plant on the microchemical environment. Summing up, it can be said that the factors influencing the distribution of epiphytes are as follows:

        (1) Age of substrate.
        (2) Rate of growth of substrate.
        (3) Light intensity.
        (4) Screening.
        (5) Nature of the surface.
        (6) Chemical surroundings.

Of these (3) is probably the most important, although it is difficult to separate its effects from those of (1) and (4).

REFERENCES

Ponds. BROWN, H. B. (1908). *Bull. Torrey Bot. Club*, **35**, 223.
Streams. BUDDE, H. (1928). *Arch. Hydrobiol. Plankt.* **19**, 433.
General. FRITSCH, F. E. (1931). *J. Ecol.* **19**, 233.
Ponds. FRITSCH, F. E. and RICH, F. (1913). *Ann. Biol. Lac.* **6**, 1.
Epiphytes. GODWARD, M. (1934). *Bei. Bot. Zbl.* **52** A, 506.
Lakes. GODWARD, M. (1937). *J. Ecol.* **25**, 496.
General. KLEBS, G. (1896). *Die Beding. der Fortpfl. ein. Algen und Pilzen.* Jena.
Ponds. LIND, E. M. (1938). *J. Ecol.* **26**, 257.
Ponds. TRANSEAU, E. N. (1913). *Trans. Amer. Micr. Soc.* **32**, 31.
General. WEST, G. S. (1916). *Algae*, p. 418. Cambridge.
Streams. FRITSCH, F. E. (1929). *New Phytol.* **28**, 165.

CHAPTER XIV

ECOLOGICAL FACTORS, GEOGRAPHICAL DISTRIBUTION, LIFE FORM

ECOLOGICAL FACTORS

Studies of the conditions controlling the distribution of algae on various rocky and salt-marsh coasts has shown that although the habitats are very different, nevertheless the controlling factors are very similar. They may be summarized briefly as follows:

(1) The nature of the coast, whether exposed or sheltered. This applies only to rocky shores because salt marshes always develop in sheltered areas.

(2) Tidal rise. This factor varies considerably from place to place, but on a rocky coast the height of the rise controls the width of the bands, the smaller the tidal rise the narrower will be the algal zones. This factor will also operate on salt marshes, but owing to the great horizontal extent of the belts it is only by accurate levelling that the effect of the factor becomes evident.

(3) Submergence and exposure operating through the daily ebb and flow of the tide. In many cases it is probable that this factor acts indirectly because it plays a considerable part in determining salinity, moisture content and water loss from the algae. There will be, however, certain species, especially the more delicate Rhodophyceae, which require to be immersed every day or which can only tolerate a few hours' exposure to drying.

(4) Non-tidal exposure, or the number of consecutive days during which no tide covers the area, is a factor which principally operates during the periods of neap tides. On the salt marshes it may assume considerable importance, especially on the higher marshes, and in such habitats it is noticed that the principal algae are either Cyanophyceae or marsh fucoids, both of which are protected against desiccation by their histological structure or by the presence of a mucilaginous envelope. Very few Chlorophyceae appear capable of withstanding long periods during which they are not covered by the tide, although they may be found in salt pans on high marshes where the presence of the water enables them to exist. Unfortunately this factor has never been studied on a rocky

coast and hence it is impossible to estimate its importance, but towards high-water mark it must operate in preventing the upward spread of a number of species.

(5) Temperature. Rees (1935) as well as Knight and Parke (1931) consider that changes of temperature throughout a season are probably responsible for the upward and downward migration of some species on the shore. Temperature would only appear to operate in the summer on high salt marshes where there is a low-growing vegetation, because it will then result in much evaporation with a consequent increase in desiccating conditions together with a rise in salinity.

(6) Salinity. This probably varies but little on a rocky coast, except perhaps in the case of tide pools, but it is important on the higher salt marshes in summer when the values rise so high in the surface layers of the soil that probably only a few algae can tolerate the conditions. The salinity of the marsh soil has also been invoked in order to explain the origin of spirality in the marsh fucoids. On all types of shore the incidence of fresh water flowing down from the land always produces a local modification in the flora.

(7) Substrate. The nature of the substrate, whether solid rock, boulders, pebbles, sand, mud or peat, is of fundamental importance in connexion with anchorage, the general aspect of the flora being largely determined by this factor. On the rocky coasts the angle of slope may affect the occurrence locally of some species, or even whole belts of vegetation (cf. p. 314).

(8) The movement of water, apart from the ebb and flow, plays a great part in determining the luxuriance of the vegetation. In many cases there may be considerable local currents and in salt marshes there is the continual flow of water in the creeks which persists even when no tide is present. The action of surf may also be included here, and there are several species, e.g. *Postelsia*, which are known to be surf-loving, whilst there are also those species which cannot tolerate surf. In places where the water carries a heavy load of silt there may be some modification in the flora because it is probable that some species are not able to tolerate consistent deposition of a muddy covering. Recent work shows that turbulence has a depressing effect upon the respiration of marine algae which is particularly marked in the case of the littoral species, and this may indirectly be related to the zonation.

(9) Biota. On a rocky coast this is largely concerned with over-shadowing and the degree of epiphytism which may often reach such proportions that the host plant is torn away with its massive burden because of the resistance offered to the water. There may also be animals, usually molluscs, which feed on the seaweeds, and these can be present in such quantity as to reduce the number of plants considerably or even to keep the area bare. An example of this is the behaviour of *Hydrobia Ulvae* on Norfolk marshes where it is present in such abundance that certain areas are kept more or less clear of *Monostroma* and *Ulva*. In addition to the molluscs there is the further problem of the phanerogamic vegetation on the salt marshes. In certain cases this may provide additional shade or lower the surface evaporation so that algae can grow at higher levels than they would do on the open marsh, e.g. *Catenella repens* around bushes of *Suaeda fruticosa* on the Norfolk marshes. The density of the phanerogamic vegetation, e.g. swards of *Puccinellia maritima* or *Spartina patens*, may prevent any real algal vegetation from developing. This can be seen on many west coast marshes of England and also on the marshes of New England.

(10) Light. Measurements show that the incident light is cut down very considerably at even a depth of 1 m., and hence algae living near low-water mark will be existing under very different light conditions to those near high-water mark. This factor is said by many workers to be of great importance in determining vertical range, but it is of course very difficult to disentangle its effect from that of the other factors. In heavily silt-laden waters this factor will probably assume even greater dimensions.

Although all these factors may be operating continually through-out the year, it must not be forgotten that only one factor operating at the critical period in the life history of a single species may be of even more importance. Johnson and Skutch (1932) have stressed this point, and they maintained that a maximum water loss during the most active growing period may be of paramount importance in determining the presence or absence of some species.

With this general introduction we may now turn to consider studies dealing more specifically with zonation on a rocky shore. Baker (1909, 1910) carried out numerous field observations on the algal zones found around the Isle of Wight, and also conducted experiments in which the four principal fucoids were grown in jars and treated artificially to different periods of exposure. As a result she came to the conclusion that the essential control of zonation was height (modified by exposure), substrate and sunshine. It is difficult, however, to see how the effects of exposure can be separated from those of actual height, and there would appear to be no good reason why exposure was not the principal determining factor. Grubb (1936) has suggested that submergence and emergence are the most important factors in determining the occurrence of algal zones, but it would appear, however, that all these factors really operate indirectly through the degree of desiccation that the different species can tolerate. Since this is bound up with their physiological economy it may be expected to have more significance than just simply height or exposure *per se*, because the real control must be related to the physiology of the plant. Gail (1920) has declared that it is the desiccation of young plants which prevents the appearance of algae outside their usual zones, and it is a remarkable fact that sporelings of fucoids are usually very strictly confined from an early stage to the zones occupied by the adult plants. As sporelings of the fucoids are not readily identified specifically when young, field experiments with young plants become extremely difficult, if not impossible, to perform. Berthold (1882) was so much impressed by the importance of this factor that he divided the rocky shore into five zones based on the degree of desiccation. It has been concluded that species growing high up on the shore have a power of resisting desiccation which is not possessed by those growing lower down, and also that those species which resist desiccation best possess the slowest growth in contrast to the others which do not resist desiccation and grow more rapidly. Fig. 202 A compares the distribution of the principal fucoids from various areas in relation to the tidal levels, and it has been suggested that the demarcation between the *Fucus spiralis* f. *platycarpus* and *Ascophyllum* zones is probably caused by desiccation, whereas the determination of the other limits may be partially or wholly explained by one or more of the following factors:

(a) Bottom structure. Boulders are essential for the attachment of *Ascophyllum* but smaller stones will suffice for the other species.

(b) Water movements, although the evidence here is somewhat conflicting.

(c) Light.

Pringsheim (1923) and Zaneveld (1937) have both shown that the water loss of the four species is very great, especially during the first 18 hours (cf. fig. 203). *Fucus spiralis* f. *platycarpus* loses its

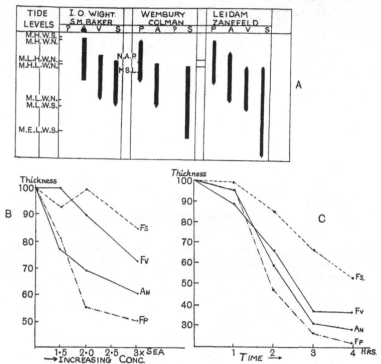

Fig. 202. A, distribution of Fucaceae on various coasts in relation to the tide levels. M.H.W.S. = mean high water mark spring tides, M.H.W.N. = mean high water mark neap tides, M.L.H.W.N. = mean low high water mark neap tides, M.H.L.W.N. = mean high low water mark neap tides, M.L.W.N. = mean low water mark neap tides, M.L.W.S. = mean low water mark spring tides, M.E.L.W.S. = mean extreme low water mark spring tides, M.S.L. = mean sea level, N.A.P. = Amsterdam tide datum line. B, decrease in diameter of cell walls when placed in sea water of increasing concentration. C, decrease in diameter of cell walls under normal conditions of exposure. An = *Ascophyllum nodosum*, Fp = *Fucus platycarpus*, Fs = *Fucus serratus*, Fv = *F. vesiculosus*. (After Zanefeld.)

water the slowest of all, and a definite increase in the rate of water loss can be observed with the different species as each occupies a successively lower zone on the shore, but it must be noted that *F. spiralis* f. *platycarpus* ultimately loses a higher percentage of water than the other three. Haas and Hill (1933) also showed that the higher the alga grows the greater is the fat content (p. 288), and hence the thickness of the cell wall must be of some significance.

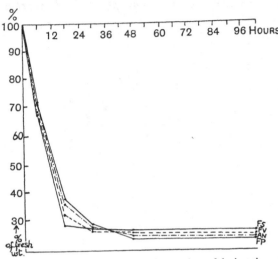

Fig. 203. Loss in weight of Fucoids in relation to time of desiccation. The higher an alga grows the slower it loses water and the greater the total loss. Symbols the same as in Fig. 202. (After Zanefeld.)

Subsequent examination has shown that the thickness of the cell wall does bear a relation to the height at which an alga grows.

TABLE XXI

	Thickness in divisions of $3\,\mu$
Fucus spiralis	0·49 ± 0·05
Ascophyllum nodosum	0·34 ± 0·01
Fucus vesiculosus	0·23 ± 0·03
Fucus serratus	0·14 ± 0·01

These cell walls decrease in thickness when subjected to desiccating conditions, and the higher a fucoid is growing on the shore the more the cell walls shrink on drying; so it must be assumed that a large part of the water lost is contained in the cell walls (cf. fig. 202 B, C). Those species which lose water most slowly will also reabsorb it most slowly and, as a result, the growth rate of the highest species will therefore tend to be the slowest. It would appear from this study that the real factor controlling zonation, so far as the fucoids are concerned, is the biochemical nature and properties of the cell wall, although it is also possible that these features have appeared as a result of the habitat they occupy.

Of those members of the Fucaceae which appear in belts, *Pelvetia canaliculata*, which forms the highest zone, is subject to the greatest exposure, but the situation of the algae in relation to each other and the density of the flora will also affect the water loss. Actual measurements carried out in the field show that the loss of weight curves for this alga are characteristically hollow, the greatest loss being in the first 6 hours, whilst the total loss may be

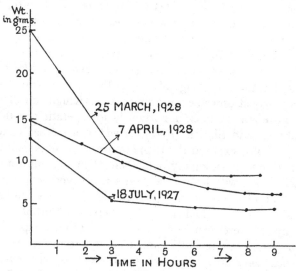

Fig. 204. Loss of water, as represented by loss in weight, in *Pelvetia canaliculata* during intertidal exposure. (After Isaac.)

between 60 and 68 % during periods of 8–9 hours (cf. fig. 204). *Fucus spiralis* f. *platycarpus* shows the same order of water loss as *Pelvetia* but then it only occupies a slightly lower belt. Besides being able to tolerate a considerable water loss which may enable it to live in a relatively inhospitable habitat, the plants of *Pelvetia*, in order to succeed, must be able to reproduce under such conditions. The two ova are not liberated from the thick-walled mucilaginous megasporangium as they are in *Fucus* and so the antherozoids must penetrate this envelope, and although this structure secures the protection of the eggs under desiccating conditions there is apparently no protection for the antherozoids. We do seem to be arriving gradually towards a state when the factors controlling the zonation of fucoids on the seashore are really becoming understood, but much more still remains to be discovered, especially in respect of the species occupying the lower belts.

Rock pools are commonly encountered on most rocky coasts, and for this reason it is perhaps desirable that they should be mentioned here. Klugh (1924) studied a series of six pools on the coast of New Brunswick and concluded that the factors which may affect the flora are (1) character of the bottom, (2) depth of the pool, (3) amount of wave action, (4) temperature, (5) salinity, (6) pH, which probably depends to a large extent on the proportion of chlorophyllous organisms in the pool, and (7) light. Many algae are tolerant of a wide pH but there is often only a narrow range in which they will develop to their greatest extent, whilst it is also possible that the percentage of iron in the liquid medium or solid substrate may at times be a limiting factor. Klugh considered that temperature was the most important factor operating in tide pools and Johnson and Skutch (1932) arrived at a similar conclusion. Whilst a pool is exposed the temperature of such a small body of water may rise considerably, and then when the tide returns the cold sea water will lower the temperature very suddenly. An examination of the flora showed that Rhodophyceae tended to be more abundant in shaded pools whereas Chlorophyceae and Phaeophyceae were relatively more abundant in the exposed pools.

Biebl (1937) has recently published the results of a study of seven rock pools on the English south coast, in which particular attention was paid to the influence of different concentrations of seawater on the cells of various species of the Rhodophyceae. As a result of this study he concluded that light, temperature, pH and seawater concentration were the important factors operating in tide pools, though they are subject to some qualification. Temperature is more likely to be effective in determining the northern limits of species rather than actually causing damage, because it was discovered that warming up to 26° C. over a period of 24 hours has no effect on most Rhodophyceae, and changes of 12° C. could occur on a hot day without causing any damage. *Spondylothamnion multifidum*, for example, apparently cannot tolerate temperatures lower than +3° C. and so reaches a northern limit on the English south coast. Most of the algae investigated tolerate a pH range of 6·8–9·6 and the pH of most pools, except perhaps the highest, will rarely be outside these values. Table XXII shows that the intertidal algae, which form the principal component of the tide pool vegetation, exhibit a greater range of osmotic tolerance than those from deep waters, although season and time of day is important in this respect because either may bring about changes in the concentration.

It would appear that in spite of all this careful work we are still far from understanding the factors that control the vegetation of tide pools because many of the algae will tolerate the range of conditions which are likely to be found in such places. The algae of the tide pools can be placed into one of the four following groups:

(1) Those which are sublittoral and which also occur in the tide pools.

(2) Those which grow near the ebb line and reach their upper limit in the pools.

(3) Those which grow in both the intertidal zone and the pools.

(4) Those confined wholly to the rock pools.

Klugh and Martin (1927) studied the growth rate of various algae in relation to submergence by measuring plants and then tying them to floats which were suspended in the water at different depths. After some months the floats were pulled up and the plants were remeasured. Light, temperature and salinity all vary with depth, but the last two factors vary so little that it is doubtful whether they can be of any significance. Light, however, is very rapidly absorbed by the water, so that at about 2 m. down only 25 % of the surface light has penetrated. The curves (fig. 205) show that maximum growth occurred between 1 and 2 m., and it would seem that whilst the light was perhaps too bright at the surface, nevertheless it soon became limiting at depths which varied for the different species. On the basis of a rather limited number of species and experiments it was concluded that the Rhodophyceae are no better adapted to greater depths than the Chlorophyceae and Phaeophyceae, but in the light of more recent experiments it would seem that this conclusion must be revised. Using coloured light under experimental conditions Montfort (1934) showed that there was an essential confirmation of Englemann's complementary theory, which states that the colour of an alga is complementary to the colour of light that it absorbs (cf. also p. 293). The phycoery-

TABLE XXII. *Osmotic resistances of Rhodophyceae from different habitats*

Algae	Diluted sea water										Concentrated sea water											Habitat and depth
	0·0	0·1	0·2	0·3	0·4	0·5	0·6	0·7	0·8	0·9	1·2	1·3	1·4	1·5	1·6	1·7	1·8	1·9	2·0	2·2	3·0	
Heterosiphonia plumosa	+	+	o	o	o	o	1	1	1	1	1	1	o	+	+	+	+	+	+	+	+	Plymouth Bay (8–10 m.)
Polyneura Hilliae	+	+	+	+	o	1	1	1	1	1	1	1	+	+	+	+	+	+	+	+	+	Wembury, tide pool
,, ,,	+	+	+	+	o	1	1	1	1	1	1	1	+	o	+	+	+	+	+	.	+	Plymouth Bay (8–10 m.)
Cryptopleura ramosa	+	+	o	o	o	o	1	1	1	1	1	.	.	.	.	.	.	.	.	.	.	,, ,,
var. *uncinatum*	+	+	o	o	1	1	1	1	1	1	1	1	1	o	o	o	+	?	+	+	+	,, ,,
Brongniartiella byssoides	+	o	o	o	1	o	1	1	1	1	1	1	1	o	o	o	o	+	+	+	+	,, ,,
Phycodrys rubens	+	o	o	o	1	1	1	1	1	1	1	1	1	1	1	o	1	o	o	o	+	,, ,,
Antithamnion tenuissimum	+	+	o	+	1	1	1	1	1	1	1	1	1	1	1	1	1	1	1	o	+	,, ,,
A. plumula	o	o	+	1	1	1	1	1	1	1	1	1	1	o	o	o	o	o	o	o	+	Tromsö, 30 cm.
A. cruciatum	o	o	1	1	1	1	1	1	1	1	1	1	1	o	o	o	o	o	o	.	+	Naples
Polysiphonia urceolata	+	o	o	1	1	1	1	1	1	1	1	1	1	1	1	1	o	o	o	?	o	Wembury, tide pool
,, ,,	+	+	o	o	1	1	1	1	1	1	1	1	1	o	o	o	o	o	o	o	?	Tromsö, L.T.M.
Membranoptera alata	+	+	o	o	1	1	1	1	1	1	1	1	1	1	1	1	1	1	1	1	o	Wembury, above L.W.M.
Ptilota plumosa	o	o	o	o	1	o	1	1	1	1	1	1	1	1	1	1	1	1	1	1	o	,, ,,
Ceramium ciliatum	+	o	o	o	1	1	1	1	1	1	1	1	1	1	1	1	1	1	1	o	o	,, ,,
Call. tetragonum var. *brachiatum*	+	o	1	1	o	1	1	1	1	1	1	1	1	1	1	1	1	1	1	?	o	Wembury, tide pool
Spondylothamnion multifidum	+	+	+	o	1	o	1	1	1	1	1	1	1	1	1	1	1	?	1	o	+	,, ,,
Griffithsia flosculosa	+	+	+	+	+	o	o	o	1	1	1	1	1	1	1	1	1	1	1	1	+	,, ,,
G. furcellata	.	.	.	.	.	.	.	.	.	.	.	.	1	o	o	o	o	o	o	o	.	Naples
G. opuntioides	.	.	.	.	.	.	.	.	1	1	1	1	1	o	o	o	o	o	.	o	.	Naples from 60 cm. up
Nitophyllum punctatum	+	+	+	o	+	o	o	o	1	1	1	1	1	o	o	o	o	o	o	+	+	Wembury, tide pool

1, all living;   o, part dead;   +, all dead.

thrin-rich red algae assimilate far better in green light than the green and brown algae, and this is valid for both surface and deep-living species. In blue-green light, on the other hand, the performance of the fucoxanthin-rich Phaeophyceae far surpasses that

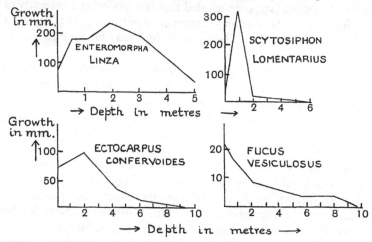

Fig. 205. Rate of growth of various algae at different depths in sea water, New Brunswick. (After Klugh.)

of the Rhodophyceae. It has also been found that the algae of all three groups can be divided into those which are shade algae and those which are sun algae, each possessing a structure that permits of maximum assimilation under the respective conditions.

TABLE XXIII

Shade algae	Sun algae
Cladophora rupestris, Chlorophyceae	Ulva lactuca, Chaetomorpha linum
Dictyota dichotoma, Phaeophyceae	Fucus serratus, Pelvetia canaliculata
Rhodymenia palmata, Rhodophyceae	Ceramium rubrum, Porphyra laciniata

Ehrke (1931), however, divided the algae into sun and shade groups on the basis of their compensation points, i.e. the light value at which respiration and assimilation are equal. On this criterion the Chlorophyceae and Phaeophyceae form one group, the sun algae, and the Rhodophyceae behave as shade algae. It is evident that the division into sun and shade algae is of considerable significance, but the basis upon which the division ought to be made would appear to necessitate further experimental work. For example, it has recently been shown that, under persistently changed conditions a fresh-water 'sun' algae can be converted into a 'shade' area.

As a result of an exhaustive study of colour in relation to assimilation, Montfort (1934) concluded that the quality and intensity of the light form the limiting factors in determining the depth at which an alga can live. These conclusions may be summarized as follows:

(1) An alga may go deeper in the water the nearer its assimilation curve approaches that of the shade type and the lower is its compensation point. (Compensation point = that strength of light in which the minimum assimilation will compensate for respiration.) The better the protoplasmic adaptation to the strong, deep-going, blue-green light waves, the greater is the power of colonizing the deeper areas. Under these conditions, for example, a green shade alga would be able to go to a lower limit than a red sun alga.

(2) An alga will go deeper the more its colour is complementary to the spectral composition of the light. Chromatic adaptation by means of Phycoerythrin and Phycocyanin may enable a red alga to have a greater energy absorption in blue-green light than a green alga, even if under conditions of white light the green alga has a greater light absorption than the red.

GEOGRAPHICAL DISTRIBUTION

Many of the studies of algal distribution are based on a consideration of continuous or discontinuous distribution which are, for convenience, discussed as though they were separate phenomena, although it is clear that no distribution can be absolutely continuous. When, however, it is found that an area in which the localities are fairly close together is separated by the width of a continent or of an ocean from another similar area, then we may talk of discontinuous distribution. The problem is rendered more difficult by the unreliability of earlier records and the somewhat scanty literature, especially for tropical and sub-tropical areas. The few studies (Svedelius, 1924; Börgesen, 1934), that have been published have established certain general features which are briefly summarized below:

(1) There is a general resemblance between the algal floras of the West Indies and the Indo-Pacific. Vicarious pairs of species

(two separate species closely related morphologically and yet widely separated geographically) are known and even vicarious generic groups. The genus *Hormothamnion* in the Cyanophyceae, *Microdictyon* and *Neomeris* in the Chlorophyceae (cf. fig. 206), all have a Caribbean-Indo-Pacific discontinuity, whilst there are several vicarious pairs in the genus *Udotea*. The explanation of these discontinuities which has been advanced by Murray, namely change of climate in former epochs, would only appear to explain certain cases, e.g. certain species in the Laminariaceae (cf. below), whilst it is equally obvious that the factors operating at present do not provide an adequate explanation. The only feasible hypothesis would be to postulate migration during an earlier epoch when there was a sea passage through the Panama isthmus, and this involves a migration not later than the Cretaceous.

(2) There are some species which are common only to the Western Atlantic and the western part of the Indian Ocean around Madagascar e.g. *Chamaedorus peniculum* and three species of *Cladocephalus*. (Cf. fig. 207.) Although there is at present no very adequate explanation for this distribution three possible hypotheses may be suggested, but there does not appear to be any evidence which supports one of them more than the others:

(*a*) Migration via the Cape.

(*b*) Migration via the Pacific and Panama, the related species perhaps still existing in the Pacific but not yet recorded.

(*c*) The related species or representatives in the interzone have died.

(3) There are some genera which are common to the Mediterranean and the Indo-Pacific region, e.g. *Codium Bursa* group, the vicarious pair *Halimeda tuna* in the Mediterranean and *H. cuneata* in the Indo-Pacific, *Acetabularia mediterranea* and other species of *Acetabularia* in the Indo-Pacific (cf. fig. 208). In this case also the only satisfactory explanation is the existence of a former sea passage across the Suez isthmus. In the flora of the northern part of the Arabian sea, out of a total of 137 species and varieties, 22 % are endemic, 52 % are Indo-Pacific and 59·6 % also occur in the Mediterranean and Atlantic Ocean, the most striking example being *Cystoclonium purpureum* which does not now exist between its widely separated stations along the southern shores of

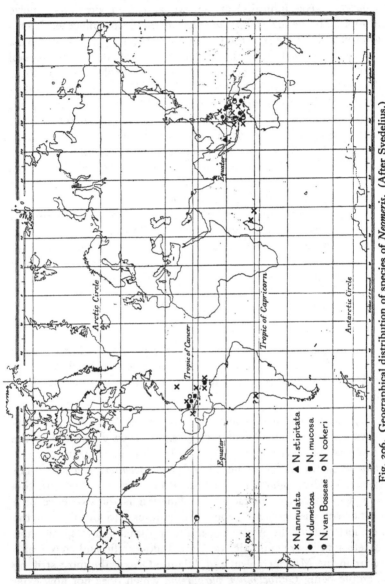

Fig. 206. Geographical distribution of species of *Neomeris*. (After Svedelius.)

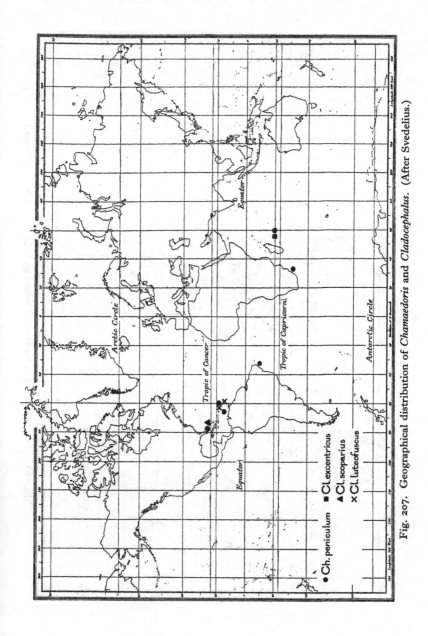

Fig. 207. Geographical distribution of *Chamaedoris* and *Cladocephalus*. (After Svedelius.)

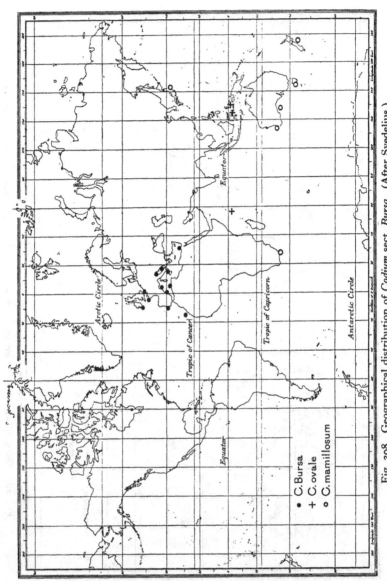

Fig. 208. Geographical distribution of *Codium* sect. *Bursa*. (After Svedelius.)

France and in the Northern portion of the Arabian Sea. In the case of the Indo-Pacific species of the Arabian sea it is often found that they are absent from the intervening tropical waters, so that their distribution must be explained as occurring at a period when the tropical waters had a more equable temperature.

(4) In general, the Indo-Pacific region is more probable as the home of the various tropical and subtropical genera and they can be classified into:

(*a*) genera with no Atlantic representatives,
(*b*) genera with a few Atlantic species, e.g. *Halimeda, Caulerpa, Sargassum, Dictyota, Scinaia, Galaxaura.*

The following genera are probably of Atlantic origin: *Dasycladus, Penicillus, Cladocephalus, Batophora.*

(5) Several families in the Laminariales, e.g. Laminariaceae, Alariaceae, are of Boreal Atlantic—Pacific discontinuity. These families must formerly have had a circumarctic distribution but were pushed south by the onset of the Ice Age and then they remained in their new habitat when the ice retreated. In this case change of climate in a former epoch provides a satisfactory explanation of the present discontinuity. Other genera, however, e.g. *Lessonia, Macrocystis, Ecklonia,* are of Antipodes-Northern Pacific discontinuity, *Macrocystis* in particular being primarily circumantarctic, after which it is absent from the tropics, to reappear again on the Pacific coast of North America and around the shores of South Africa. The two species of the southern hemisphere appear to be identical with the two species in the northern hemisphere so that presumably they have disappeared from the intervening warm zone. Again, it must be concluded that their migration took place at a time when the temperatures of the ocean waters were more equable, unless it is assumed that the species have since become less tolerant towards temperature.

Apart from these facts of general distribution there is very little further information in the literature. The Danish workers, Börgesen and Jonsson (1905) and Jonsson (1912), have studied the arctic and subarctic floras in some detail and their results may properly be included here. They concluded that the component species of the flora could be divided into a number of distinct groups:

(1) The arctic group, with its southern European border in north Norway and Iceland, although in America the group may extend as far south as Cape Cod.

(2) A subarctic group, the species of which are common in the Arctic sea and the cold boreal area of the Atlantic as far south as western France.

(3) Boreal arctic group. These species are common in the Arctic Sea and the boreal area of the Atlantic as far south as the Atlantic coast of North Africa, some perhaps penetrating even farther south.

(4) A cold boreal group which is of more limited distribution, extending northwards from western France to south Iceland and Finland, with outlying species penetrating in the south to the Mediterranean and in the north to the White Sea and Sea of Murman.

(5) A warm boreal group, the species of which extend as far south as the Mediterranean and Atlantic coast of North America, some perhaps even farther south. Their northern limits are to be found in south Iceland, the Faeroes, north-west Norway and Scotland.

Although Iceland is so far north, nevertheless the flora is predominantly boreal because 54 % belong to the last three groups. If the different districts of Iceland are compared with neighbouring floras it is extremely interesting to see how the floras of the various parts of the Icelandic coast show resemblances to floras from a number of widely separated areas.

TABLE XXIV

	Groups 1 and 2	Groups 3-5
East Greenland	81	19
Spitzbergen	77	23
West Greenland	72	28
East Iceland	63	37
Finland	46	54
South-west Iceland	42	58
South Iceland	50	70
Faeroes	29	71
Nordhaven	27	73

Another interesting feature in geographical distribution, which has been established by Setchell (1920), is the relation of the various species to the isotherms. The surface waters of the Oceans are divided into zones according to the courses of the 10°, 15°, 20° and 25° C. isotheres. The great majority of algal species are confined to only one zone, a considerable number occur in two, only a small number occur in three zones, whilst the number extending over four or five zones are very few indeed and their distribution is usually by no means certain. In New England many of the species are apparently separated by the 20° C. isothere which approximates closely to the position of Cape Cod, so that the flora to the north of the Cape is essentially different to that of the south. Those species limited to one zone are called *stenothermal* whilst the wide ranging forms are termed *eurythermal*. The former species are particularly characteristic of the warmer waters, but, even so, many apparent eurythermal species are found on examination to be essentially stenothermal. *Monostroma Grevillei* and *Polysiphonia urceolata* are summer annuals in the cold waters of Greenland, but in the southern part of their range they develop in winter and early spring when the temperature will be the same as it is in the Greenland summer. With the exception, then, of the temperatures endured by the resting spores they are essentially stenothermal. *Ascophyllum nodosum*, with a temperature range from 0 to 10° C., is another case and in the southern part of its range the plants pass into a heat rigor during the hotter months.

Feldmann (1937) has recently drawn attention to the phenomenon of seasonal alternation of generations and seasonal dimorphism. In *Ceramium corticatulum* the tetrasporic plants exist only at the end of autumn or in the winter whilst the sexual plants are to be found at the end of summer. This is an example of seasonal alternation of generations in which there are ephemeral summer haploid plants with the diploid plants occurring during the winter and persisting over a longer period. Seasonal dimorphism is exhibited in the Mediterranean by *Cutleria multifida* and *C. monoica* with their sporophytes *Aglaozonia parvula* and *A. chilosa*. The two species are almost indistinguishable morphologically, but the former occurs in spring in shallow waters off-shore whilst the latter occurs in summer at greater depths. Another example of seasonal dimorphism is shown by the two morphologically similar species *Polysiphonia sertularioides* and *P. tenerrima*, the former occurring on exposed rocks from December to May whilst the latter grows epiphytically on *Nemalion helminthoides* between June and December.

A word may conveniently be said here about the behaviour of some species in relation to fish and fisheries (Savage, 1932). One of the most outstanding examples is *Phaeocystis pouchetii*, a coloured flagellate which, when present in quantity, gives the waters of the North Sea a muddy appearance, the so-called "baccy juice". Herrings are repelled by this organism when it is present in mass, and the vernal maximum of this organism off the Dutch coast turns the northward herring migration west towards the coast of E. Anglia and thus brings about the spring fishery (cf. fig. 209A, B). The occurrence of an abnormal autumn maximum out of its usual station may completely change the grounds of the autumn fishery during the southward migration: such an abnormal maximum is known to have occurred in 1927 (cf. fig. 209C).

LIFE FORM

A study of algal ecology leads one to the conclusion that the distribution of the different types appears to be largely controlled by the nature of the habitat, e.g. rocky shore, sandy shore or salt marsh, although of course there may be other factors because this will not explain the predominance of the large kelps in the colder waters and the predominance of the lime-encrusted forms in the warmer waters. For this reason there would seem to be a need for some sort of Life Form classification comparable to that of Raunkiaer's for the flowering plants. Such a system can be used to give a quantitative picture of the composition of the vegetation and also to demonstrate the absence of any type, thus raising the problem as to why they are absent. Biological spectra, similar to those employed by Raunkiaer (1905), form a convenient way, if used with caution, of comparing floras from two different areas although they are subject to the limitation that they do not indicate the dominant types.

Oltmanns' schema of 1905, which is one of the earliest, is based largely upon morphological criteria, but in the light of present knowledge it is more desirable to adopt a scheme with some relation to habitat rather than one based on purely morphological characters:

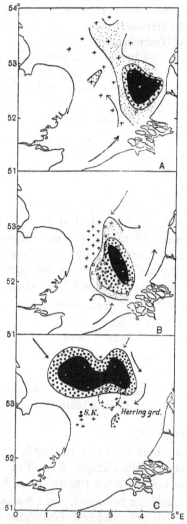

Fig. 209. *Phaeocystis* and herrings. A, distribution of *Phaeocystis*, 17–24 April, 1924, normal distribution. +, *Phaeocystis* scarce or absent. ○, stations in *Phaeocystis* zone. Intensity of concentration shown by shading. → assumed herring migrations. B, distribution of *Phaeocystis*, 8–13 April, 1926. Spring fishery interference. C, distribution of *Phaeocystis*, 6–9 November, 1927. Autumn fishery interference. *S.K.* = Smith's Knoll Lightship. (After Savage.)

(1) Bush and tree forms (*Bryopsis*).
(2) Gelatinous bush forms (*Diatoms*).
(3) Whip forms (*Himanthalia*).
(4) Net forms (*Hydrodictyon*).
(5) Leafy forms: (*a*) lattice (*Agarum*),
　　　　　　　　(*b*) flag (*Macrocystis*),
　　　　　　　　(*c*) buoy (*Nereocystis*).
(6) Sack forms (*Leathesia*).
(7) Dorsiventral forms (*Delesseria*).
(8) Cushion, disk and encrusting forms (*Ralfsia*).
(9) Epiphytes, endophytes and parasites.
(10) Plankton.
(11) Symbionts.

In 1927 Funk proposed a new classification which applied particularly to the algae of the Gulf of Naples. He distinguished four primary groups, all of which were capable of subdivision according to the same principles, but unfortunately the terms that he employed for the major groups are not particularly happy as some of them are open to the widest interpretation:

I. Seaweeds ("Tange" in the original).
II. Lime-encrusted algae.
III. Fine algae ("Feinalgen", or algae of small proportions).
IV. Microscopic algae, including species measuring less than 1 cm.

Each of these groups could be subdivided as follows, the examples being taken in this case from the first group.

I. *Sea weeds* ("*Tange*"):
(*a*) Large algae, more than 1 m. in length, e.g. *Laminaria*.
(*b*) Medium algae, with a length of 0·5–1 m., e.g. *Fucus*.
(*c*) Small algae ranging from 1 to 50 cm. in length:
(i) Main axis not branched, e.g. *Chaetomorpha*.
(ii) Main axis branched, e.g. *Gracilaria*.
(iii) Thallus bushy, e.g. *Gelidium*.
(iv) Thallus leafy or a foliose bush, e.g. *Phyllitis*.
(v) Creeping thallus, e.g. *Caulerpa*.
(vi) Crustaceous thallus, e.g. *Ralfsia*.
(vii) Thallus a hollow ball, e.g. *Colpomenia*.

Gislen in 1930 proposed another classification to include both plants and animals, the biological types referable to the plants being as follows:

I. CRUSTIDA (Crustaceous thallus):

    (1) Encrustida or encrusting forms, e.g. *Lithothamnion*.
    (2) Torida or small cushions, e.g. *Rivularia*.

II. CORALLIDA (lime skeleton more or less developed):

    (1) Dendrida or tree-like forms, e.g. *Corallina*.
    (2) Phyllida or leaf-like forms, e.g. *Udotea*.
    (3) Umbraculida or umbrella-like forms, e.g. *Acetabularia*.

III. SILVIDA (no lime skeleton):

    (*a*) Magnosilvida, or forms more than 1 dcm. high and with branches more than 1 mm. thick.

        (1) Graminida, e.g. *Zostera* (a phanerogamic group).
        (2) Foliida, e.g. *Laminaria*.
        (3) Sack-form, e.g. *Enteromorpha*.
        (4) Palm form, e.g. *Lessonia*.
        (5) Buoy form, e.g. *Nereocystis*.
        (6) Cord form, e.g. *Himanthalia*.
        (7) Shrub-like form, e.g. *Chordaria*.
        (8) *Sargassum* form.
        (9) *Caulerpa* form.

    (*b*) Parvosilvida (small delicate forms less than 1 dcm. high).

It will be seen that all these classifications are based primarily upon morphological criteria and are therefore incomplete because they do not take into consideration the biological requirements of the algae.

Setchell propounded a scheme in 1926 based primarily on the conditions found in tropical waters, with particular reference to coral reefs. For this reason the classification is restricted because it would require considerable extension if the flora of colder waters were to be included, but at the same time it is an improvement over the previous schemas in that its basis is largely ecological:

24-2

HELIOPHOBES:

(1) *Pholadophytes.* Forms nestling into hollows and avoiding much light.

(2) *Skiarophytes.* Forms growing under rocks or in their shade.

HELIOPHILES:

(3) *Metarrheophytes* or attached flexible forms growing in moving water.

(4) *Lepyrodophytes* or encrusting forms.

(5) *Herpophytes* composed of small creeping algae.

(6) *Tranophytes* or boring species.

(7) *Cumatophytes* or "surf-loving" species.

(8) *Chordophytes*, where the thallus has the form of a cord.

(9) *Lithakophytes* or lime-encrusted species (Corallinaceae).

(10) *Epiphytes.*

(11) *Endophytes.*

Knight and Parke (1931) proposed a brief classification based upon the same criteria, duration and perennation, that Raunkiaer employed for the higher plants. They only distinguished four groups; perennials, pseudoperennials, annuals and casual annuals, and it would require a thorough restudy of many species in order to determine to which group they belong. More recently (1937) Feldmann has proposed a new scheme, based on these same criteria, which can be regarded as the logical elaboration of Knight and Parke's classification:

(1) *ANNUALS*

(a) Species found throughout the year. Spores or oospores germinate immediately.

      EPHEMEROPHYCEAE: *Cladophora.*

(b) Species found during one part of the year only.
  (i) Algae present during the rest of the year as a microscopic thallus.

      ECLIPSIOPHYCEAE: (a) with prothallus, *Sporochnus.*
                    (b) with plethysmothallus, *Asperococcus.*

  (ii) Algae passing the unfavourable season in a resting stage.

      HYPNOPHYCEAE—Resting stage:

            (a) spores, *Spongomorpha lanosa.*
            (b) oospores, *Vaucheria.*
            (c) hormogones, *Rivularia.*
            (d) akinetes, *Ulothrix pseudoflacca.*
            (e) spores germinate and then become quiescent, *Dudresnaya.*
            (f) protonema, *Porphyra.*

(2) *PERENNIALS*

(a) Frond entire throughout year.
  (i) Frond erect. PHANEROPHYCEAE: *Codium*.
  (ii) Frond a crust. CHAMAEPHYCEAE: *Hildenbrandtia*.

(b) Only a portion of the frond persisting the whole year.
  (i) Part of the erect frond disappears. HEMIPHANEROPHYCEAE: *Cystoseira*.
  (ii) Basal portion of thallus persists.

HEMICRYPTOPHYCEAE:
  (a) basal portion a disk, *Cladostephus*.
  (b) basal portion composed of creeping filaments, *Acetabularia*.

This scheme must be regarded as a great advance on the other classifications, but at the same time it does not seem to take adequate account of the effect of environment and, furthermore, it is primarily of use for the marine algae and does not take into consideration the numerous fresh-water and terrestrial species.

Cedergren (1939) has recently published a life form scheme based primarily upon the nature of the medium and secondarily upon the nature of the substrate. This scheme can be considered as excellent in so far as it classifies the algae in a more general sense.

Series A. *Air-loving Algae.*

(1) Terricolae (on the earth).
(2) (a) Saxicolae (on stone).
  (b) Calcicolae (on chalk).
(3) Lignicolae (on wood).
(4) (a) Epiphytes.
  (b) Endophytes.
(5) Epizoic forms (on animals).
(6) Succicolae (gelatinous).

Series B. *Soil Algae* (in the earth).

Series C. *Water Algae.*

(1) Nereider (river and stream algae).
(2) Limnaeider (lake algae).
(3) (a) Epiphytes.
  (b) Endophytes.
(4) (a) Epizoic forms.
  (b) Endozoic forms.
(5) Plankton (small floating algae).
(6) Pleuston (large floating algae).
(7) Neuston.

Of all those so far published, however, Feldmann's appears to be the most workable. The real test will come if and when it is employed to give biological spectra, and if the spectra from different localities e.g. temperate and tropical regions, show a distinct

374   ECOLOGICAL FACTORS, ETC.

difference then it should prove possible to extend its use as a means of comparing the vegetation from different regions. Such differences may be expected to open up problems, the solutions of which should yield us valuable information concerning the general biology and ecology of the species concerned.

Ecology. BAKER, S. M. (1909, 1910). *New Phytol.* **8**, 196; **9**, 54.
Ecology. BERTHOLD, G. (1882). *Mitt. Zool. Staz. Neapel,* **3**.
Ecology. BIEBL, R. (1937). *Beih. bot. Zbl.* **57** A, 381.
Geographical Distribution. BÖRGESEN, F. (1934). *Det. Kgl. Danske Vidensk. Selsk. Biol. Meddel.* **11**, 1.
Geographical Distribution. BÖRGESEN, F. and JONSSON, H. (1905). *Botany of the Faeroes,* **3**. Copenhagen.
Life Form. CEDERGREN, G. R. (1939). *Bot. Notiser,* p. 97.
Ecology. CHAPMAN, V. J. (1937). *J. Linn. Soc. (Bot.),* **51**, 205.
Ecology. EHRKE, G. (1931). *Planta,* **13**, 221.
Life Form. FELDMANN, J. (1937). *Rev. Alg.* **10**, 1.
Life Form. FUNK, G. (1927). *Publ. della Staz. Zool. Napoli,* **7**.
Ecology. GAIL, F. W. (1920). *Publ. Puget Sd Biol. Sta.* **2**.
Life Form. GISLEN, T. (1930). *Skr. K. Svensk Vetensk.* nos. 3, 4.
Ecology. GRUBB, V. M. (1936). *J. Ecol.* **24**, 392.
Ecology. ISAAC, W. E. (1933). *Ann. Bot., Lond.,* **47**, 343.
Ecology. JOHNSON, D. S. and SKUTCH, A. F. (1928). *Ecology,* **9**, 307.
Geographical Distribution. JONSSON, H. (1912). *Botany of Iceland,* Part I, p. 58. Copenhagen.
Ecology. KLUGH, A. B. (1924). *Ecology,* **5**, 192.
Ecology. KLUGH, A. B. and MARTIN, J. R. (1927). *Ecology,* **8**, 221.
Ecology. KNIGHT, M. and PARKE, M. (1931). *Manx Algae,* p. 20. Liverpool.
Ecology. MONTFORT, C. (1934). *Jb. wiss. Bot.* **79**, 493.
Life Form. OLTMANNS, F. (1905). *Morphologie und Biologie der Algen,* **2**, 276. Jena.
Ecology. PRINGSHEIM, E. G. (1923). *Jb. wiss. Bot.* **62**, 244.
Ecology. REES, T. K. (1935). *J. Ecol.* **23**, 69.
Life Form. RAUNKIAER, C. (1905). *Acad. Roy. Sci. Let. Dan.* **5**, 347.
Geographical Distribution. SAVAGE, R. E. (1932). *J. Ecol.* **20**, 326.
Geographical Distribution. SETCHELL, W. A. (1920). *Amer. Nat.* **54**, 385.
Life Form. SETCHELL, W. A. (1926). *Univ. Calif. Publ. Bot.* **12**, 29.
Geographical Distribution. SVEDELIUS, N. (1924). *Arch. Bot.* **19**, 1.
Ecology. ZANEFELD, J. S. (1937). *J. Ecol.* **25**, 431.

INDEX

Numbers in heavy type refer to the figures

Geosiphon, 296
Germany, 271
GETMAN, M. R., 211
GIBB, D. C., 211, 320
Giffordia secundus, 133
Gigartina, 306, 307, 310, 315
Gigartinaceae, 239, 240
Gigartinales, 214, 238
Girvanella, 267
GISLEN, T., 374
Gleucocystis, 295
Gloeocapsa, 7, 12, 17, 262, 267, 296; *crepidinum*, 7
Gloeocapsomorpha, 266
Gloeochaete, 295
Gloeocystis, 9
Gloeodinum, 126, 264; *montanum*, **125**
Gloeothece, 9, 267
Glyceria, 338
Glycogen, 1, 6
Gobia, **151**
GODWARD, M., 341, 345, 348
GOEBEL, K., 126
GOMONT, M., 298
Gomphonema, 347
Gongrosira, **67**, 95, 263
Gonidia, 7, 13, 32
Gonimoblast(s), 214, 220, 222, 223, 225, 227, 233, 236, 237
Goniolithon, 273
Gonium, **24**, 36; *pectorale*, **24**, 25
Gracilaria, 370; *confervoides*, 238
GRAEBNER, P., 298
Graminida, 371
Greenland, 299, 366, 367
GRIFFITHS, MRS, 233
Griffithsia, 215, 233, 244; *corallina*, 215, 233, **234**; *flosculosa*, 358; *furcellata*, 358; *globulifera*, 233; *opuntioides*, 358
GROSS, F., 126
GROSS, I., 56
GROVES, J., 126
GRUBB, V. M., 215, 244, 320, 352, 374
GRINTZESCO, J., 44
Gunnera, 300
GUSSEWA, K., 62
Gymodinium, 264; *aeruginosum*, **125**
Gymnosolen, **267**
Gyrogonites, 273

HAAS, P., 288, 304, 353
Haematochrome, 18, 19, 33
Haematococcus, 32, 36; *pluvialis*, **33**

HAINES, H., 302, 304
Halarachnion ligulatum, 291, 292
Halicystaceae, 86
Halicystis, 73, 86, 87, 97, 166, 261; *ovalis*, **86**, 87, 251
Halidrys, 158, 203, 306, 311; *dioica*, 203; *siliquosa*, 203, **204**, 289
Halimeda, **93**, 97, 269, 276, 277, 365; *cuneata*, 361; *tuna*, 361
Halopteris filicina, 247
Halosaccion, 265, 317
Halosphaera, 115, 126; *viridis*, **116**
Halosphaeraceae, 115
HAMEL, G., 136, 154
HAMMERLING, J., 84
HANSON, E. K., 281, 304
Hantschia amphroxys, 299
Hapalosiphon arboreus, 9
Haplobionts, 213, 243, 250
Haplonts, 213, 250
Haplospora, 160; *globosa*, **161**, 162
Haplostichineae, 127
Haptera, 129
Harpacticus chelifer, 168
Harpenden, 338, 341
HARPER, R. A., 36, 44
HARTMANN, M., 20, 36, 51, 56
HARVEY, G., 238
Harveyella, 216, **238**, 239, 244; *mirabilis*, 238
Haustoria, 239
Hawaii, 219
HEILBRON, I. M., 5
Heligoland, 291
Heliophiles, 372
Heliophobes, 372
Hemicryptophyceae, 373
Hemiphanerophyceae, 273
Herpophytes, 372
Herrings, 368
Heterocapsaceae, 114
Heterochloridaceae, 114
Heterochloridales, 114
Heterochloris, 264
Heterococcales, 114
Heterocyst(s), 8, 14, 15, 16
Heterogeneratae, 127, 128, 131, 132, 155, 167, 189, 190, 260
Heterokontae, 1, 2, 18, 98
Heterosiphonales, 114, 118
Heterosiphonia plumosa, 358
Heterotrichales, 114, 262
Heterotrichy, 254, 255, 262, 277
HIGGINS, E. M., 162
HILDENBRANDT, F. E., 262

Printed in the United States
By Bookmasters